# Would-Be Worlds

# Would-Be Worlds

## How Simulation Is Changing the Frontiers of Science

### John L. Casti

**John Wiley & Sons, Inc.**
New York ■ Chichester ■ Brisbane ■
Toronto ■ Singapore ■ Weinheim

Library of Congress Cataloging-in-Publication Data

Casti, J. L.
    Would-be worlds : how simulation is changing
the frontiers of science / by John L. Casti.
        p.    cm.
    Includes bibliographical references and index.
    ISBN 0-471-12308-0 (cloth : alk. paper)
    1. Computer simulation.    I. Title.
QA76.9.C65C38   1996
003'.7—dc20                                96-16841

Printed in the United States of America
10 9 8 7 6 5 4 3 2 1

# CREDITS

Grateful acknowledgment is made to the following for permission to reproduce material used in creating figures in this book. Every reasonable effort has been made to contact the copyright holders of the material used here. Omissions brought to our attention will be corrected in future editions.

Figures 1.1 and 1.2, Frontpage Sports: *Football Pro'95* Screen Shots © Sierra-On-Line, Inc. and used with permission.

Figure 1.3, Reprinted with permission from Academic Press, Inc., Orlando, Florida, from S. Dole, *Icarus,* volume 13, pp. 500–501. Copyright © 1970 by Academic Press.

Figures 1.4 and 1.5, Crystal, David, *The Cambridge Encyclopedia of Language,* copyright © 1987 by Cambridge University Press. Reprinted with the permission of Cambridge University Press.

Figure 1.7, Figure 31 from *Myths, Models, and Paradigms* by Ian G. Barbour. Copyright © 1974 by Ian G. Barbour. Reprinted by permission of HarperCollins Publishers, Inc.

Figure 1.8, The Metropolitan Museum of Art, Bequest of Gertrude Stein, 1946. (47.106)

Figure 1.9, by permission of the Stadtische Mannheim, Germany.

Figures 2.1 and 2.2, Dawkins, Richard, *The Blind Watchmaker,* Longman Pub. Co., copyright © 1986 by Richard Dawkins. Reprinted with permission of the author.

Figure 2.3, Copyright © Clifford Pickover, *Computers, Patterns, Chaos, and Beauty,* Clifford Pickover. Reprinted with permission of St. Martin's Press, Incorporated.

Figure 2.5, From P. Prusinkiewicz and A. Lindenmayer, *The Algorithmic Beauty of Plants,* Springer, 1990. Reprinted with permission of the author.

Figure 2.6, P. Prunsinkiewicz, M. Hammel, and E. Mjoiness. "Development of Hieracium umbellatum" in *Proceedings of Siggraph '93.* © Copyright 1993 Association for Computing Machinery. Reprinted by permission.

Figure 2.7, From T. Saaty and P. Kainen, *The Four Color Problem,* © 1986, Dover Publications Inc. Reprinted with permission of Dover Publications, Inc.

Figures 2.8 and Figure 2.9, From Eigen M., *Steps Toward Life,* © 1992, Oxford University Press, Oxford, UK. Reprinted with permission of Oxford University Press.

Figure 2.10, From Henderson-Sellers A. and K. McGuffie, *A Climate Modelling Primer,* © 1987. Reprinted by permission of John Wiley & Sons, Ltd.

Figure 2.11, From *Laws of the Game* by Manfred Eigen and Ruth Winkler, trans. R&R Kimber. Translation copyright © 1981 by Alfred A. Knopf Inc. Reprinted by permission of the publisher.

Figure 2.12, From *Digital Mantras: The Languages of Abstract and Virtual Worlds,* by Steven R. Holtzman, MIT Press, 1995.

Figure 2.13, © 1991, Association of Computing Machinery. Reprinted by permission.

Figure 2.14, © 1991, Association of Computing Machinery. Reprinted by permission.

Figure 2.15, © 1991, Association of Computing Machinery. Reprinted by permission.

Figure 2.16, Courtesy of Christopher Crawford.

Figure 2.17, Courtesy of Brian Arthur and John Holland.

Figure 3.1, Reprinted with permission from *Dynamical Systems—A Renewal of Mechanism,* p. 3, 1986. © 1986 World Scientific Publishing Co.

Figure 3.2, From *The Paradoxicon,* by N. Faletta, © 1983. Reprinted with permission of Bantam Doubleday Dell.

Figure 3.3, From *New Scientist,* October 3, 1992, p. 14. Courtesy *New Scientist.*

Figure 3.4, *American Scientist,* Jan.–Feb.1995, "The Organization of an Ant Colony." © 1995. Reprinted with permission of Sigma XI.

Figures 3.6 and 3.7, Jackson, E. Atlee, *Perspectives of Nonlinear Dynamics,* Vol. 2, Fig. 7.79, p. 205, "Stretching and Folding"; Figure 7.95, p. 216, "A saltwater taffy-pulling machine." Copyright © 1990 by Cambridge University Press. Reprinted with the permission of Cambridge University Press.

Figures 3.8, 3.9, and 3.10, From N. Hall, ed., *The New Scientist Guide to Chaos,* figs. 6.1, 6.2, and 6.3, 1992. Reprinted with permission of Timothy Palmer.

Figure 3.11, Courtesy of Brian Arthur.

Figure 3.12, Courtesy of Stuart Kauffman.

Figure 3.13, From R. Atkin, *Mathematical Structure in Human Affairs,* fig. 3.1, page 47. Copyright © 1974 by R. Atkin. Used with permission of the author.

Figure 3.14, From *Fractals, Chaos, and Power Laws* by M. Schroeder. Copyright © 1991 by W. H. Freeman and Company. Used with permission.

Figure 3.15, Courtesy of William Bowers.

Figure 3.16, From F. Flam, "Hints of a Language in Junk DNA," *Science,* 266, November 25, 1994, p. 1320. Reprinted with permission of Eugene Stanley.

Figures 4.2–4.9, Courtesy of Chris Barrett.

Figure 4.10, Courtesy of A. K. Dewdney.

Figure 4.11, From Jackson, P., *Introduction to Artificial Intelligence,* 1974, fig. 1.2, p. 13.

Figure 4.12, From Eccles, J., *Facing Reality,* p. 10, fig. 3A, New York: Springer-Verlag, Inc., 1970. Reproduced with permission of Springer-Verlag.

Figures 4.13, 4.14, and 4.16, "Introduction to Neural Networks," California Scientific Software, Nevada City, CA, © 1994.

Figure 4.15, From Arbib, M., *Brains, Machine, and Mathematics,* 2ed., p. 18, fig. 2.3a, New York: Springer-Verlag, Inc. 1987. Reproduced with permission of Springer-Verlag.

Figure 4.17, Courtesy of Melanie Mitchell.

Figure 4.18, Courtesy of Tom Ray.

Figures 4.19–4.21, Courtesy of Joshua Epstein and Rob Axtell.

Figure 5.1, From W. Brock et. al, *Nonlinear Dynamics, Chaos, and Instability* (1991). Copyright © 1991 by MIT Press. Reprinted with the permission of MIT.

Figure 5.4, Thompson, Sir D'Arcy, *On Growth and Form,* "Two species of fish," Fig. 5.19 and 5.20, p. 1062, Copyright © 1942 by Cambridge University Press. Reprinted with the permission of Cambridge University Press.

Figure 5.5, Courtesy of J. O'Connor.

Figures 5.6 and 5.7, From *The Armchair Universe* by K. Dewdney. Copyright © 1988 by W. H. Freeman and Company. Used with permission.

# Contents

# Preface

When one of my friends saw the original manuscript of this book, he remarked, "Oh, boy. You finally wrote a book with no equations, Greek symbols, or any other kind of mathematics. What happened? Did you slip up? Or did you just decide to take pity on us mathematical illiterati?" Out of politeness, or more likely to take the intellectual high ground, I probably replied to the effect that it's not necessary to fly off into the mathematical stratosphere in order to explain the strange ways of complex systems. That may be what I said; but it's certainly not what I was thinking. What was going through my mind was that it's really a pity that this book is *not* crammed full of mathematical arcana, since if it were it could only mean that we had something that looked like a decent *theory* of complex systems. We don't. In fact, we're not even close. *Would-Be Worlds* is a book about why we're not even close, what the nature is of the barriers standing in the way of such a theory, and how researchers today are working to surmount these barriers.

Complex systems pervade every nook and cranny of daily life, from the patterns of traffic in urban transport networks to the movement of prices on financial markets. These processes are fundamentally different from the simple systems that have constituted the focus of the scientific enterprise since the time of Newton. Simple systems generally involve a small number of individual elements with relatively weak interactions between them, or they are systems like enclosed gases or distant galaxies, composed of such vast numbers of objects that we can employ statistical averaging techniques to study their behavior. Complex systems, on the other hand, involve a medium-sized number of agents: drivers, traders, molecules .... Moreover, in complex systems the agents are generally both intelligent and adaptive, in the sense that they make decisions in accordance with various rules, and are ready to modify their rules of action on the basis of new information that comes their way. There are no dictators or centralized controllers in these systems; no single driver, trader, or molecule has access to what everyone else in the system is doing, so the agents in a complex system make their decisions and update their rules for action on the basis of *local,* rather than global, information.

These properties of complex systems—a medium-sized number of intelligent, adaptive agents acting on the basis of local information— differ so greatly from the simple systems that science has studied up to now that I feel it's safe to say their investigation represents an entirely new chapter in the pursuit of scientific knowledge of the world around us. Fortuitously, just at the moment in history when the unforeseen behavior of these systems is bringing them most forcefully to our attention, technology has presented us with a marvelous tool with which to probe their mercurial nature. This tool, of course, is the digital computer.

One of the most revered principles in the philosophy of science is the so-called scientific method, by which one arrives at scientific knowledge of real-world phenomena. An integral part of this method is the idea of controlled, repeatable experiments to test hypotheses about how the world can be the way it is. Without a laboratory in which to perform these kinds of experiments, there can be no such thing as a bona fide *scientific* theory of anything. Such laboratories have taken many forms: the chemical lab with its complement of test tubes, retorts, and Bunsen burners, multibillion-dollar particle accelerators for probing the heart of matter; and the everyday high-school biology lab, in which we all carved up frogs and looked through microscopes at plant cells. But these are labs devoted to the study of the *material* structure of simple systems. Perhaps more than anything else, *Would-Be Worlds* is devoted to an account of the computer as a laboratory for studying the *informational* structure of complex systems.

The creation of silicon surrogates of real-world complex systems allows us to perform controlled, repeatable experiments on the real McCoy. So, in this sense complex-system theorists are in much the same position that physicists were in at the time of Galileo, who was responsible for ushering in the idea of such experimentation on simple systems. It was Galileo's efforts that paved the way for Newton's development of a theory of such processes. Unfortunately, complex systems are still awaiting their Newton. But with our newfound ability to create worlds for all occasions inside the computer, we can play myriad sorts of what-if games with genuine complex systems. No longer do we have to break the system into simpler subsystems or avoid experimentation completely because the experiments are too costly, too impractical, or just plain too dangerous. *Would-Be Worlds* gives a first-hand account of this informational revolution, which in my opinion will form the basis for so-called normal science in the coming century.

As the reader makes his or her way through this volume, it won't take long to discover the strong influence on this book of work done at the Santa Fe Institute (SFI). This is no accident. If someone were to ask 50 years from now what SFI ever contributed to the good of humankind and/or the advancement of science, my guess is that the answer will be that SFI's place in the scientific firmament will be identified with pioneering the idea of using the computer as a laboratory for the study of complex, adaptive systems—especially those systems arising in the social, biological, and behavioral sciences. Because this work is still in its infancy, it's completely unremarkable that recent work at or around SFI shines through on almost every page of the book. In this regard, I'd like to take this opportunity to thank my many colleagues at SFI who have given so generously of their time helping to make sure that what I'm saying here is not too far off the wall. So in no particular order, a tip of my hat for services rendered to Chris Langton, Brian Arthur, Nelson Minar, Blake LeBaron, Buz Brock, Stu Kauffman, Tom Ray, Anders Karlqvist, Murray Gell-Mann, Gustav Feichtinger, Rob Axtell, Martin Shubik, Chris Barrett, Peter Schuster, Mike Simmons, Doyne Farmer, Steen Rasmussen, Jim Gardner, Melanie Mitchell, Josh Epstein, Joe Traub, Walter Fontana, and Harold Morowitz.

I would also like to express my gratitude to Professor Åke Andersson of the Institute for Future Studies in Stockholm. The stimulating intellectual environment of the institute, as well as its partial financial support of this project, has been instrumental in shaping much of the content of this volume. In this same regard, I am grateful to Ms. Hanna Alen, who suggested the book's title during a very pleasant coffee-time discussion at the institute.

In addition, special thanks to Atlee Jackson and David Lane, both of whom read the entire manuscript and made many crucially important suggestions that materially improved both the content and the presentation. Of course, the usual accolades to my editor, Emily Loose, who so strongly supported the idea of the book from the start. Her insightful (and inciteful!) comments, coupled with her insistence that the material be comprehensible to a layperson, helped turn a dry, dusty, academic tome into something that I hope is readable by just about anyone. If it is not, blame the author.

JLC
Santa Fe, New Mexico

# CHAPTER

# 1

# Reality Bytes

## Up and Down the Electronic Gridiron

Contrary to popular belief, the world's greatest sporting event in terms of prolonged, worldwide interest is not the Olympic Games. Rather, it is the World Cup of football, which, like the Olympics, is held just once every four years and is played out over a period of two weeks or more. The United States hosted this global spectacle in 1994—and what a spectacle it turned out to be. The championship game between Brazil and Italy was witnessed live by a crowd of over 100,000 people in the Rose Bowl in Pasadena, California, and by a crowd of at least a billion on television the world over. This grand finale on July 17 was finally won 3 to 2 by Brazil on penalty kicks, after the two teams had struggled through 120 minutes of stirring but scoreless play.

    With two such evenly matched clubs as Brazil and Italy, I think many soccer fans would wonder along with me about how the two teams would have fared had they the opportunity to play each other again— or even 10 or 100 times more? Would the Brazilians have emerged as the clearly superior club in such an extended round of competition? Or would the two teams have battled on a more-or-less even basis, as one might infer from the outcome of the one actual game played that fateful day in Pasadena? I guess we will never really know. But one way of at least imagining how this question might be answered is to create a computer simulation of the two teams. We could then let them fight it out on the electronic field of play instead of the grass and dirt of the Rose Bowl. To be convincing, such a simulation would have to take into account the playing characteristics of each and every player and the strategies of the Brazilian and Italian coaches, not to mention random factors like field condition, wind speeds, home field advantage, crowd noise, and all the other minutiae that enter into determining the outcome of most sporting events.

Unfortunately, I know of no such simulations for World Cup teams. But there is just such a product available for playing American professional football inside your computer. So let's have a look at this very type of experiment with what is perhaps the world's third most popular sporting event, the Super Bowl football game, which settles the championship of American professional football each January. For those readers unfamiliar with American football, all you need to know about the game for the discussion that follows is that it is very similar to soccer in both form and content: It involves two teams competing to score the most points by putting the ball beyond their opponent's goal line by running with the ball, throwing it, or kicking it. So with this in mind, let's turn to Super Sunday and the showdown between the San Francisco 49ers and the San Diego Chargers, which settled the National Football League title for the 1994–1995 season.

## Super Sunday

On Sunday, January 29, 1995, the San Francisco 49ers met the San Diego Chargers at Miami's Joe Robbie Stadium in Super Bowl XXIX. As half the world knows by now, the 49ers crushed the Chargers by a count of 49 to 26 in what was certainly one of the most boring Super Bowls ever played, the Chargers acting as if they knew right at the outset that they didn't belong on the same field as the 49ers by yielding two quick touchdowns before the game was even four minutes old. What a laugh!

As I sat watching this miserable contest unfold in the comfort of my den, I couldn't help but wonder what the outcome would be if these two clubs were to meet each other in a long series of games instead of this single real-world encounter. In such an extended sequence of plays, would one team or the other demonstrate the kind of clear superiority that the 49ers showed in their one-time-only encounter in Miami? Or would the series of outcomes look more like the sequence you'd get from just flipping a fair coin, tending to an approximately even split in wins and losses between the two teams? What if the games were played not on a neutral field like Joe Robbie Stadium, but half were played at each club's home field in the kind of weather that prevails in San Francisco and San Diego in late January? To carry this fantasy onto an even higher plane, suppose you didn't like the game plan created by 49ers' coach George Seifert or Charger headman Bobby Ross, believing you could do a better job of coaching these teams yourself. Could you really do better?

It's armchair quarterbacking and what-if questions like these that keep football fans busy during the long off-season from February through August. What gives these kinds of questions their particular charm and pizzazz is that they lie forever in the realm of idle speculation; there is just no way to answer them by actually doing the experiment. Or is there? Is it even remotely possible that modern technology can provide us with a way to test such fantastic speculations objectively in the laboratory rather than relegate them to the unanswerable?

Until just recently, the answer to this query would have been a quite clear-cut No way! A couple of years ago, however, computer-game developers released football simulations that effectively enable fans to create the entire National Football League (NFL) inside their computing machines. Probably the most sophisticated and detailed of these games is *Football Pro '95 (FBPRO95)*, published by a firm called Sierra On-Line, Inc. In this highly detailed simulation, every player on every team in the NFL is represented and rated in categories such as speed (SP), acceleration (AC), agility (AG), strength (ST), hands (HA), endurance (EN), intelligence (IN), and discipline (DI). These ratings are determined by formulas involving individual and team statistics appropriate for the player's position. Table 1.1 shows these ratings for some of the NFL's leading quarterbacks at the end of the 1994 season. Putting these individual ratings together with customizable game plans, team and coaching profiles, and factors such as weather conditions at the league's outdoor stadiums, the computer gamester is effectively able to create a *laboratory* with which to experiment with the NFL.

**Table 1.1**    1994 NFL quarterback ratings in *FBPRO95*.

| Name | SP | AC | AG | ST | EN | HA | IN | DI |
|------|----|----|----|----|----|----|----|----|
| Troy Aikman (Dallas) | 72 | 73 | 69 | 78 | 54 | 75 | 85 | 90 |
| Steve Young (San Francisco) | 74 | 84 | 83 | 75 | 69 | 79 | 86 | 89 |
| Dan Marino (Miami) | 57 | 60 | 69 | 84 | 69 | 65 | 88 | 87 |
| Brett Favre (Green Bay) | 72 | 83 | 84 | 74 | 62 | 62 | 84 | 79 |
| Stan Humphries (San Diego) | 71 | 76 | 83 | 82 | 80 | 66 | 80 | 88 |

To see how experiments in this football laboratory work, let's go back to Super Bowl XXIX. In real life, we know that the game was won by the San Francisco 49ers over the San Diego Chargers by a score of 49 to 26. To see if this was just a fluke, I sprinkled these two teams onto the electronic gridiron provided by *FBPRO95* and let them fight it out ten times. Figure 1.1 shows the image on my screen as the two teams lined up to begin a play during one of these 10 tussles. Along with the teams on the field of play, the figure also shows the game clock in the lower-left corner indicating the time remaining in the quarter and the play clock in the lower-right corner displaying the number of seconds left for the team with the ball to put it into play. In Figure 1.2 we see the final scoreboard for game 10, which was won by San Francisco by a score of 21 to 7. The scoreboard also shows the highest-rated plays for this particular stage of the game, as given by each team's game plan for this particular encounter. But these are not important for our discussion here. The overall results of the 10 simulated Super Bowl XXIXs are shown in Table 1.2, where we see that the two teams split the series 50–50 in this surrogate world, each winning five of the 10 games. In terms of total points scored, the 49ers scored 187 points and the Chargers scored 182, a skimpy per-game advantage for San Francisco of only half a point, differing rather markedly from the actual margin of victory of 23 points. But points and games won or lost are not the only way to test the accuracy or utility of the laboratory experiment. Here's another.

For those not averse to placing their money where their mouths are, a small wager on the Super Bowl adds a touch of piquancy to the game, as well as providing a very tangible reason to root for one team over the other. As the Las Vegas sports books listed the game, the 49ers were a 19-point favorite. For those unfamiliar with sports-betting jargon and practice, what this means is that to win a bet placed on San Francisco,

**Figure 1.1**  Play during a simulated run of Super Bowl XXIX (see Plate I).

the 49ers would have to win the game by more than 19 points; if not, the bookies pay off to those backing the Chargers. If the final score differential were to be a San Francisco victory by exactly 19 points, then all bets are off and no money changes hands.

From this bettor's perspective, the 10 simulated Super Bowls look quite different than they do from the fan's viewpoint of mere wins and

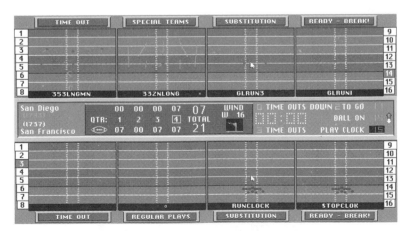

**Figure 1.2**  Final scoreboard for simulated game 10 of Super Bowl XXIX (see Plate II).

**Table 1.2**  Results of simulating Super Bowl XXIX 10 times.

| | Game Number | | | | | | | | | |
|---|---|---|---|---|---|---|---|---|---|---|
| | **1** | **2** | **3** | **4** | **5** | **6** | **7** | **8** | **9** | **10** |
| **San Diego** | 13 | 35 | 10 | 27 | 12 | 16 | 23 | 19 | 20 | 7 |
| **San Francisco** | 26 | 42 | 13 | 13 | 0 | 22 | 17 | 16 | 17 | 21 |
| **Point differential** | −13 | −7 | −3 | 14 | 12 | −6 | 6 | 3 | 3 | −14 |

losses. Looking at the point differentials listed in the last row of Table 1.2, we find that the 49ers never covered the Las Vegas line (that is, won by more than 19 points), losing against the point spread every single time. So if one were foolish enough to take these experiments on the electronic gridiron seriously, the only way to bet would have been to take San Diegoand put those 19 points into the bank. One could easily argue that 10 simulated games is not a large enough sample for the many vagaries of fate influencing the outcome of a football game to even out, and that a much larger set of experiments is needed before placing any faith (or money!) on the results. So just for fun, I set one of my computers loose for about a week replaying Super Bowl XXIX 100 times. The result was 54 victories for the 49ers with an average margin of victory of 6.67 points and 46 wins for San Diego, their average winning margin being 7.15 points.

Further statistical analysis of these results suggests that the 49ers are indeed statistically superior to the Chargers—but not by anything even remotely approximating the 19-point bookmaker's spread. In fact, in the entire set of 100 simulated games, there was *not one single game* in which the 49ers margin of victory exceeded this seemingly absurd point spread. We will explore this hard-to-explain (and hard-to-take) gap between Miami gridiron realities and its electronic counterpart in my computer shortly.

*FBPRO95* is an example of the kind of surrogate world that aims to explain a high-level phenomenon (the final score of a football game) by appealing to interactions among lower-level agents (the players and coaches of the two competing teams). Note that throughout this volume we shall employ the term *agent* to indicate the basic unit of action in our simulations. The reader may wish to substitute alternative terms, such as *decision maker, object,* or *player.* They are synonymous as far as our treatment here is concerned. Such microworlds or bottom-up models rely upon our knowledge of the agents and their interactions with each other, which then generates a pattern at a level higher than the agents themselves. Such a pattern is often termed an *emergent* phenomenon, since it emerges out of the aggregate interactions among the individual agents in the system. Our story in this volume will be about the construction of such microworlds in a computer's memory bank, the behavior of such models, and the degree to which we can employ them as electronic surrogates for the slice of reality they purport to represent. The questions we will consider include

- How do these simulated worlds relate to their real-world counterparts?

- What types of real-world systems lend themselves to these electronic surrogates and why?

- How do we build such worlds?

- How can we use these electronic worlds to study the real world?

- What types of behaviors can emerge from the interaction of agents in their electronic environment?

Let me hasten to note that the idea of building such computational worlds is not an especially new one. A cursory examination of the simulation and modeling literature of the 1950s turns up many papers suggesting such models, and even a few attempts to actually construct them. For example, many of the war-gaming exercises at The RAND Corporation, the National Defense College, and elsewhere in this period, as well as business simulation games of the time, focused on individual decision makers and their interactions as determinants of the behavior of both countries and corporations.

The obstacle, of course, is that in order for interesting phenomena to emerge, it's usually necessary to dig quite deep into the system to find the right agents to generate the behavior that interests us. It is also a challenge to describe the interactions among these agents, each of whom can learn from past behavior and thereby change the mode of their interaction with other agents as the process unfolds. In football, for instance, the players certainly do not always run the same play when facing a particular situation. If Troy Aikman of the Dallas Cowboys always threw a short look-in pass to his tight end Jay Novacek whenever it was third down and five yards to go, teams around the league would immediately jump on this pattern in the Cowboy's game plan and arrange their defensive alignment accordingly. Similarly, traders in financial markets can't always use a strategy like "buy if the market has moved up for the past three days; otherwise, sell." They have to change their strategies for buying and selling in the light of what all the other traders are doing at that moment, if they want to remain solvent.

These illustrations suggest that to build and use electronic versions of worlds like the NFL or the New York Stock Exchange effectively, we have to (a) have several agents interacting with each other, and (b) give these agents the ability to learn and modify their strategies as the process unfolds. Satisfaction of these criteria, in turn, calls for the kind of computational capabilities that researchers in the 1950s could only dream about. In fact, even as sophisticated a program as *FBPRO95* has no capabilities for the players to learn and modify their actions on the basis of what they may learn during the course of a game. The program is only capable of a simpler type of calculation. It just weights the likelihood of success of various possible actions and chooses one at random, biasing its choice by the weighting. No doubt this is due to the need to keep the computational burden of the program within the capabilities of a home computer so that fans with even modest hardware at their disposal can enjoy the game. After all, the game's publisher is in business to sell the program, not just to provide a research tool for analyzing the NFL. Even if the game's publisher did want to make *FBPRO95* more sophisticated, there is the far-from-simple matter of trying to decide what kind of learning process(es) might be at work in something like a football game.

Because of these technological and psychological reasons, modelers have tended to focus their efforts on what we might call aggregated, or high-level, models. These types of models predict patterns like the final score of a football game not from first principles, like the decisions of individual players and coaches, but from higher-level properties like yards gained passing and rushing by the two teams, their respective number of turnovers (fumbles, interceptions, and blocked kicks), and third-down efficiency. Clearly, these quantities are determined by the micro-level interaction of individual players, which means that a quantity like total yards gained can be regarded as an emergent phenomenon itself arising from the lower-level interaction among the players. Such quantities exist at a higher, more aggregated level than that of the individual players, but at a level lower than that of the total number of points scored by each team. In such models, instead of following the activities of individual agents, we combine medium-, or *meso*-level quantities, like yardage gained and turnovers, by various statistical techniques to come up with a projection of how many points these factors contribute to a team's overall total. Investors will recognize the same approach at work in estimating the price of a stock or the price of a commodities contract on frozen orange juice by statistically massaging meso-level quantities like corporate earnings or the anticipated weather next month in Florida.

Even though our treatment here is centered upon the bottom-up, microworld type of model as a way of getting at the scheme of things in natural and human systems, meso-level models should not be dismissed as merely poor cousins to the finer resolution surrogate worlds provided by the microworlds. To begin with, the higher-level, coarse-resolution models have the inestimable advantage of being computationally implementable on rather primitive computational hardware (which, of course, is a principal reason why they have dominated most computational modeling efforts until rather recently). Moreover, such models often serve admirably to answer the kinds of questions we want to know about in a given setting. For example, if all you care about is whether the price of a share of IBM stock is going to go up or down tomorrow, it is probably not necessary to have a detailed model of every trader in the market, their individual psychologies, and their current trading strategies in order to make a pretty good estimate of where the price is going. All you usually need is aggregate information about IBM's quarterly earnings report, the expected actions of the Federal Reserve on interest rates, and some feel for recent market trends.

So despite the fact that very little is said about these kinds of meso-level modeling efforts in the pages that follow, the reader faced with having to construct a model of a given situation should not immediately jump to the conclusion that a microworld is the only—or even the better—way to go. It's merely *one* way that can *sometimes* shed useful light on a situation that can't be obtained through a more coarse-grained look at the system. Having put this caveat on the record, the following section talks a bit more about models and their relationship to the real world.

## Distant Suns and Planetary Nebula

Whether they are on the micro, meso, or macro scale, models are constructed with the purpose of representing some aspect(s) of reality. However, just as with a piece of Impressionist art, which may represent reality in a fashion quite different from, say, a photograph or an engraving, various types of models may also portray the same slice of reality quite differently. To pin down some of these general ideas about model types and what can be learned from models, let's look at the very concrete problem of modeling the formation of a planetary system revolving about a central star. This will serve as an example of modeling—both good and bad—in action, and will help us get a feeling for the kinds of features that separate a good, that is, useful, model from a pretender.

If one takes the conventional wisdom in the philosophy of science seriously, the process of modeling planetary formation begins with observation. Astronomers gaze into the heavens, seeing vast gaseous clouds (planetary nebulae) swirling around in distant galaxies. These clouds often contract under gravitational forces into dense agglomerations of matter that are believed to ultimately end up as a central blob we call a star, together with smaller blobs rotating around the star that constitute the star's attendant planetary system.

After collecting a number of observations of such clouds of gas, planetologists then proceed to construct patterns, or empirical rules, linking the mass of the clouds, their rate of rotation around a central star (or stars), and the number and size of the blobs that break off from the clouds as the clouds spin faster and faster. The observed relationships between the composition and the properties of these clouds constitute the starting point for a theory of planetary formation. Such a scientific theory traditionally has to be tested against observation and refined by means of controlled, repeatable laboratory experiment. But because the direct observation of such protoplanets is extremely difficult, due to the faint luminosity of these objects compared to that of their parent sun (or suns, in a multiple-star system), investigators cannot generally test these theories by either observation or laboratory experimentation. What to do?

One of the things that can be done is to create an electronic version of the swirling gas cloud inside a computer, making use of the known laws of physics involving gravitational attraction of particles, angular momentum, and conservation of energy. In this fashion, one can introduce lumps of matter into smooth, homogeneous clouds of various compositions that are rotating in different ways, then sit back and watch what kinds of planetary formations arise. One of the first exercises of this sort was carried out a number of years ago by astrophysicist Stephen Dole of The RAND Corporation.

A sample of Dole's results are shown in Figure 1.3, with our own solar system indicated as well, for the sake of comparison. The planetary masses are expressed in units relative to the mass of the Earth (which is taken to be 1.0), and the distances from the central star are measured in astronomical units (AU), 1AU being the average distance between the Earth and the Sun. The numbers at the left of the simulated planetary systems are just labels for cases in which different amounts of matter were injected into the originally homogeneous gaseous cloud to get the condensation process going.

What is striking about these simulations is the strong similarity of the fake planetary systems to our own solar system, at least qualitatively. There appears to be a pronounced tendency for these gaseous clouds to separate out into systems consisting of a number of smaller inner planets, together with a few outer gas giants. This overall picture seems to persist in the face of a wide range of randomly scattered inhomogeneities in the initial cloud. This provides strong circumstantial evidence in support of the idea that planetary formation is a rather common phenomenon wherever one finds rotating clouds of gas in the universe.

What is the model here? Basically, the model consists of the description given to compute the physical properties of the matter forming the gaseous cloud and the interactions of these particles of matter with each other via the laws of physics and (probably) chemistry. These relations are then coded into a language that a computing machine can understand—Fortran, C++, or Pascal, for instance. At this point, the actual physical universe of gases and planets has been mapped, or coded, into a quite different universe, the world of computational objects. When we actually run our computer program and watch the electronic planetary systems of Figure 1.3 emerge, it is natural to wonder about the degree to which these systems bear any resemblance to what we might expect to find out there in the real universe. In short, how does the model match up to reality? This question, in turn, is but one aspect of a much broader issue that we shall encounter on almost every page of this work, namely, how good is the model? To answer this basic query, we have to look more closely at the modeling process itself.

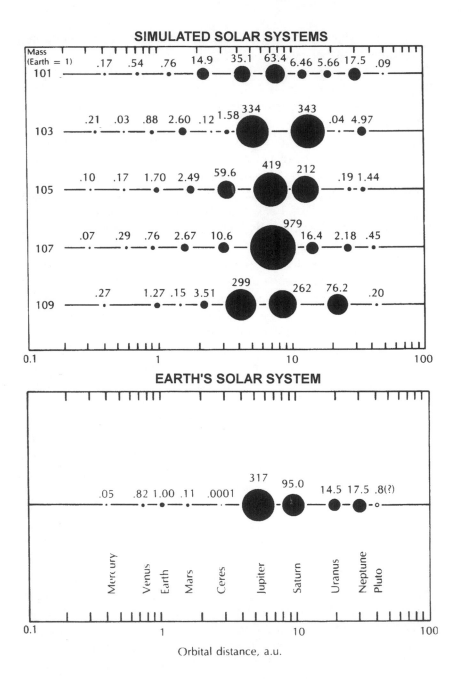

**Figure 1.3** Simulated planetary systems.

# A Gallery of Models

To those who spend their working day measuring flow patterns over airplane wings in a wind tunnel or, perhaps, thumbing through the pages of a fashion magazine like *Vogue,* the foregoing use of the term *model* may well come as a bit of a surprise. To such folk, a model is either a miniature version of some real-world object, like an airplane fuselage, or an idealized version of a real-world person, like you or me. In either case, the model has nothing to do with mathematical symbols or lines of computer code.

Models like these are what we might label *experimental* models, because they constitute material representations of reality that have either had some real-world features abstracted away (as with a model car or a ship in a bottle) or they have had imperfect real-world features, like broken noses or bowed legs, replaced by idealized versions (such as the perfect proportions of a fashion model). In either case, the model then serves to answer by direct experiment certain types of questions, such as how air flows over an airplane wing under different atmospheric conditions or how particular types of clothes (that no real person could ever afford or would ever wear) look when draped over a perfect shape. However, as important as such experimental models are in both daily and scientific life, they are not our concern here. Our interest is in models constructed out of pure information, not matter.

Turning to models based on symbolic rather than physical representations of reality, there are several varieties and flavors that must be distinguished. For example, in mathematics there is an area of research called model theory, which deals with what we might call *logical* models. To illustrate, suppose we have a collection of axioms and rules of deductive inference that allow us to start with an axiom and create new, logically correct statements from the axiom by applying the rules of the formal, logical system. At this level, both the axioms and the statements following from the rules of the deductive system are just strings of abstract symbols, a completely syntactic system. What model theorists attempt to do is find concrete mathematical systems whose rules of operation faithfully mirror the behavior of the given abstract axiomatic framework. In other words, they attempt to attach *meaning* to the rules of the formal, syntactic structure. For example, the set of points and lines in high-school geometry serves as a concrete model for the formal axioms of Euclid's geometry. The model theorist then uses this specific system of points and lines both to illustrate the formal axiomatic system and to give an interpretation of it. Again, our interest here is not with the theory of models in this model-theoretic sense.

Residing somewhere between the nitty-gritty experimental models of the wind tunnel and the mannequin's runway, the austere, abstract world of logical models is the primary focus of our attention here, the *mathematical cum computational* model. The quintessential example of such a model is Newton's famous equation of particle motion $F = ma$, expressing the acceleration $a$ of a particle as a relationship between the force $F$ acting upon it and the particle's mass $m$. Such a model may be given explicitly in mathematical terms, as when we write down a set of differential equations whose solution tells us how the position and velocity of a Newtonian particle changes over the course of time. It may be given implicitly by coding the relationship among the variables in Newton's equation into symbols and rules constituting a computer program. In either case, the model is a representation of a *physical system* using abstract symbols like $F$, $m$, and $a$.

Finally, there is the *theoretical* model. This is an imagined mechanism, or process, invented by the scientist to account for observed phenomena. Such a model is usually put forth by way of analogy with familiar mechanisms or processes, and has as its objective the understanding of the phenomena it claims to represent. A good example is Niels Bohr's famous solar-system model of the atom, in which he postulated the analogy between the atomic nucleus and the Sun, and regarded the orbiting electrons in the atom as analogous to the planets orbiting around a central star.

Like a mathematical model, such a model is also a symbolic representation of some real-world situation, but one that differs in its objective, which is to explain rather than predict. Basically, this means we want the model to account for *past* observations rather than to predict *future* ones. A theoretical model can often turn into a mathematical one simply by expressing the theoretical model in mathematical and/or computational terms. Our earlier example of planetary formation is a good example of this. There we began with a theoretical model of planetary formation involving an image of particles in interaction through gravitational attraction coalescing into a planetary system. By coding these objects and their interactions into a computer program—itself a formal mathematical structure—we created a mathematical model of the process of planetary formation. This model, in turn, allowed us to make predictions about the kinds of planetary systems that might be seen in the real universe under various physical circumstances surrounding the mass of the gaseous cloud, its composition, rate of rotation and so forth.

To summarize this brief foray into the taxonomy of models, we can distinguish at least four model types: experimental, logical, mathematical/computational, and theoretical. Our main concern here will be with the latter two, especially the computational, using these types of models to pry loose the secrets of how patterns like the outcome of a football game, the movement of equity prices, and the emergence of biological form can arise out of the actions of individual agents like football players, traders, and cells.

## Models for All Occasions

Just as there are different types of models, each having its own characteristic features, advantages, and disadvantages, there are also different ways that models can be used. Consideration of these uses gets us a bit closer to understanding what distinguishes a good model from a bad one. In the next three subsections, we present a very-far-from exhaustive account of reasons we construct models, which is a key point in separating the good from the bad and the ugly.

### Predictive Models

Probably the most famous model of all time is Newton's mathematical description of the motion of a set of particles interacting through gravitational forces. This model provides a picture of how planets move, showing why, if two bodies attract each other with a force inversely proportional to the square of the distance between them, the orbits traced out by the bodies will be elliptical in shape. Thus, it is a good explanation of the reason planets move as they do. Even more important for fishermen, farmers, and astrologers is the fact that Newton's model of planetary motion offers a procedure by which to calculate exactly where each planet will be at any given time in the future, provided we can measure the planets position and the rate and direction of motion today.

Newton's model for the motion of gravitating bodies is an example of what is called a *predictive* model. Such a model enables us to predict what a system's behavior will be like in the future on the basis of the properties of the system's components and their current behavior. The well-known *ideal gas law* is another model of this sort. This law states that the relationship between the pressure $P$, the temperature $T$ and the volume $V$ of a gas in an enclosed container is given by $PV = nRT$, where $n$ and $R$ are constants characterizing the particular type of gas under consideration. Thus, using this mathematical model of the behavior of an enclosed gas, we can confidently predict that if we raise the temperature, the pressure of the gas in the container will rise if we keep the volume fixed. In fact, we can even do better than this qualitative prediction by giving a rather accurate estimate of exactly *how much* the pressure will change for a given increase in temperature.

## Explanatory Models

The diagram shown in Figure 1.4 is a tree diagram showing the Indo-European family of languages and their geographical distributions. It indicates how the various modern European, Slavic, and Indian languages originated from an ancient language called Proto Indo-European and spread out, west to the European continent and south to the Indian subcontinent.

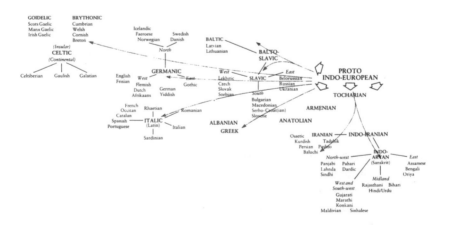

**Figure 1.4**  The Indo-European language tree.

The nineteenth-century German linguist August Schleicher originated the idea of organizing languages in this type of tree diagram. Schleicher regarded language as an organism that could grow and decay, and whose changes could be analyzed using the methods of the natural sciences. The tree diagram provides a model of this process, one that emerges from work by linguists in identifying formal similarities and differences between various languages. From this information, they then try to work out an earlier stage of language development from which all the forms could have arisen. Figure 1.5 shows how this process works using the various words for *father* in the Romance languages at the bottom of the figure. We see how they all derived from the Latin word *pater.* Even if Latin no longer existed, it would still be possible to reconstruct a great deal of its form by comparing large numbers of words in this way. Exactly the same reasoning is used when the parent language no longer exists, as when the forms in Latin, Greek, Sanskrit, Welsh, and so on are compared to reconstruct the Indo-European form given at the top of Figure 1.5. Incidentally, the asterisk next to this form indicates that it is a reconstructed form that has not been attested to in written records.

The model of Indo-European language development represented by the tree diagram of Figure 1.4 is an example of an *explanatory* model. The primary purpose of such a model is not to predict the future behavior of a system, but rather to provide a framework in which past observations can be understood as part of an overall process. Probably the most

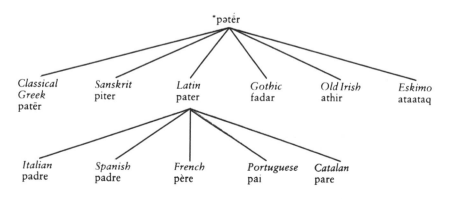

**Figure 1.5** Tree diagram for the word *father.*

famous model of this type is Darwin's Principle of Natural Selection, by which one can explain the appearance and disappearance of the many types of living things that have populated the Earth over the past four billion years or so. Both the language tree and Darwin's principle serve well to explain what has been observed about languages and organisms by providing an overarching structure into which we can comfortably fit many known facts. However, when it comes to predicting the appearance of a new word or dialect or the emergence or disappearance of a species of organism, these models remain silent. We saw earlier with Newton's equations that an explanatory model can sometimes be predictive as well. This is usually not the case, especially with models of evolutionary processes as we will see in detail later on in this book.

### Prescriptive Models

Both the predictive and explanatory models we have considered thus far have been pretty passive sorts of things, telling us about some slice of reality but not really giving us any special insight into how to perhaps shape that reality to our own ends. A *prescriptive* model remedies this shortcoming, offering us a picture of the real world that has various "knobs" built-in that we can twist one way or another to see how that reality might be bent to our will. Such models, while standard fare in economics and engineering, are rather foreign to the biologist and physicist. So let's look at an example to fix the idea.

Figure 1.6 shows a block diagram of the ebb and flow of labor, money, and materials, all of which taken together constitute a toy version of the economic life of a nation. This diagram shows, for instance, that everything else being fixed, an increase in unemployment leads to an increase in the size of the employable population, while an upward movement of incomes generates a decrease in unemployment. So the signed links in this model of the economy provide us with a method by which to change aspects of the nation's economy that we don't like. For instance, consider that perennial politician's nightmare, unemployment. If unemployment grows to an uncomfortable level, the model suggests that all we need to do is increase the manufacture of export goods somehow, which will then increase income. This, in turn, will lead to a decrease in the number of unemployed. Those familiar with economic processes will recognize that the links in this diagram have been given their signs

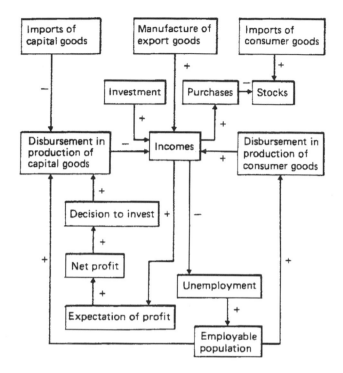

**Figure 1.6**  Model of a national economy.

in a manner that mirrors a Keynsian view of economics, in which the government plays a central role in keeping things like unemployment, manufacturing, and incomes in a always shaky sort of balance.

What makes the model of Figure 1.6 prescriptive is that it offers an explicit way for decision makers to change the behavior of the economy by intervening in different fashions. We just saw one of these ways when an increase in export goods led to a decrease in unemployment. But there are many others that the reader can easily discern by a simple inspection of the model. The point is that the model has been constructed with such intervention in mind, which makes it differ in fundamental ways from either a predictive model, like that of the Ideal Gas Law, or an explanatory model, like that of the Indo-European languages, neither of which contain explicit "handles" like levels of purchases or amount of investment that can be manipulated so as to modify the reality the model trys to represent.

From the foregoing considerations we see that not only are there different types of models, but that models may be constructed for very different purposes. Both of these factors enter into the question of how we should evaluate the performance of a model. Let us finally turn to this most central of modeling issues.

## The Philosopher's Stones

In his short story "The Seventh Sally," Polish science-fiction writer Stanislaw Lem tells the tale of the robot constructor Trurl, who builds a miniature kingdom for the evil tyrant Excelsius, the Tartarian, ruler of Pancreon and Cyspenderora, who had been driven from his throne by the inhabitants of his kingdom and exiled to a barren asteroid. Upon discovering Trurl's identity, Excelsius immediately demands that Trurl restore him to his throne, a request that Trurl has no intention of granting. But feeling sorry for the deposed monarch, Trurl proceeds to construct a scaled-down—but perfect—replica of Pancreon and Cyspenderora, complete with

> plenty of towns, rivers, mountains, forests, and brooks, a sky with clouds, armies full of derring-do, citadels, castles, and ladies' chambers; and there were marketplaces, gaudy and gleaming in the sun, days of back-breaking labor, nights full of dancing and song until dawn, and the gay clatter of swordplay . . . a fabulous capital, all in marble and alabaster, and a council of hoary sages, and winter palaces and summer villas, plots, conspirators, false witnesses, nurses, informers, teams of magnificent steeds, and plumes waving crimson in the wind . . . .

Putting this miniature kingdom into a box that could be carried around with ease, Trurl gave it over to Excelsius to rule and have dominion over forever.

The story then takes a turn to the philosophical when Trurl brags to his fellow constructor Klapaucius about his skill in handling the delicate business of safeguarding the democratic aspirations of Excelsius's former subjects, while indulging the autocratic tendencies of the deposed monarch. But to Trurl's great surprise, Klapaucius takes him to task for building this miniature civilization, arguing that such a microworld is just as real as our own civilization, and that by constructing such a world Trurl has done a gross disservice to the microinhabitants, who are now subject to the cruel whims of Excelsius. "Sheer sophistry," responds Trurl, asserting that these so-called inhabitants are merely the minuscule caperings of electrons in space, arranged by his inestimable skill as a constructor and programmed to behave in a fashion mimicking down to the finest detail their real-world counterparts on Pancreon and Cyspenderora.

So what do we have here? Is Trurl's microkingdom a good model of Pancreon and Cyspenderora or, as Klapaucius claims, is it *too* perfect to be good? These are the kinds of questions that lead us back to the consideration of what constitutes a good model, and how we identify these elusive characteristics.

## Semper Fidelis?

The second part of the title of Lem's story about the microkingdom of Excelsius is "Or, How Trurl's Perfection Led to No Good," strongly suggesting that mere duplication of the real world in a model, in this case what we earlier termed an experimental model, is in no way an indication that the microworld is a good model of its macroworld counterpart. In fact, most of Klapaucius's objections to Trurl's wizardry revolve about the fact that Trurl's creations are just a bit too perfect, somehow acquiring the same degree of reality as their macroworld counterparts simply by virtue of their perfection. Therefore, complete fidelity in the model is certainly far from a sufficient condition to deem the model a good representation of reality.

Perfect fidelity is not a necessary condition for a model to be good, either. For example, let's look at one of the most well known and useful models in the kinetic theory of gases, the so-called *billiard-ball model.* This model regards the individual molecules of an enclosed gas as being little billiard balls, flying about inside the container and crashing into each other in an unimaginably large number of elastic collisions. The overall picture is one of a *very large* number of billiard balls moving about in three-dimensional space, creating the aggregate properties of the gas—pressure, temperature, and volume—through their collisions and interactions with the walls of the container. The general structure of this model (which is what we earlier termed a theoretical model) is shown in Figure 1.7.

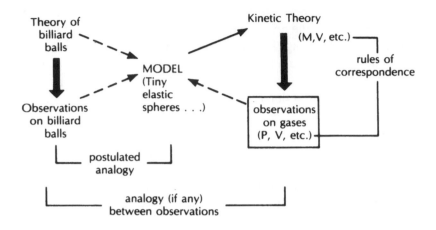

**Figure 1.7**  The billiard-ball model of a gas.

The billiard-ball model has been enormously successful as a mechanical picture with which to understand many aspects of the behavior of a gas, and has served to motivate the development of both statistical mechanics and the kinetic theory of gases. However, even a passing look at the components of the model show that it is sheer fiction as far as providing an accurate representation of *real* gas molecules and their interactions. Real molecules do not bounce off each other in completely elastic collisions. Nor are such molecules the hard, microscopic, point particles of the billiard-ball type. So the model is far removed from being a macroworld duplication of what's actually going on in the microworld of the gas. Note here that the scaling goes in the opposite direction from that in the world of King Excelsius, now proceeding from the micro to the macro. We could just as well regard the enclosed gas as being a model for the movement of billiard balls, instead of the other way around. It doesn't really matter in which direction the scaling goes; what counts is the fidelity of that direction we regard as being the model. In neither case can we argue that a duplication of the real-world situation in the model is a necessary condition for the model to be thought of as being good.

This observation calls to mind a remark made by Pablo Picasso, in response to complaints that his portrait of Gertrude Stein (Figure 1.8) didn't look at all like her. Picasso replied, "Everybody thinks she is not at all like her portrait, but never mind. In the end she will manage to look just like it." In fact, in later years the portrait was indeed acclaimed

**Figure 1.8**  Picasso's portrait of Gertrude Stein.

as being an admirable likeness of the writer. If we were to think of Picasso's portrait as a model of Gertrude Stein, then what was seen as the reality in some sense *was* the model. Shades of Einstein's widely quoted statement to the effect that, the theory (read: model) tells you what you can observe.

We come then to the surprising conclusion that it is neither necessary nor sufficient for a good model to faithfully capture all aspects of the phenomenon it represents. What kind of standards can we conjure up that do serve to separate the good models from the bad?

## The Art of the Model

On several occasions we have referred to a scientific model of a real-world actuality as a *picture* of that phenomenon. So perhaps we can push this metaphor just a bit further as a way of getting just a bit closer to pinning down what we mean by a "good" model.

Suppose we want to inform ourselves about the looks of Gertrude Stein. Will we learn more from Picasso's portrait of her or from a photograph? Which will be the better representation, the portrait or the photograph? Consider another even more graphic example of what's involved in coming to terms with this question.

The Emperor Maximilian was executed in Mexico in 1867, an event that was depicted in Edouard Manet's painting *The Execution of*

*Maximilian* shown below in Figure 1.9. Manet based his painting on eyewitness accounts reported in newspapers and on photographs of portraits of Maximilian and his military staff. Even though Manet's artistic account of this event is not literally a faithful depiction of the actual execution itself, it has been said to reveal the true synthesis of the impersonal forces that resulted in Maximilian's death and the sympathy and admiration that Maximilian elicited in France.

Picasso's portrait and Manet's painting get us closer to one of the most characteristic features of a good model: Such a model captures the *essence* of its subject. Roughly speaking, this means that the symbols and objects of the model are sufficiently rich to allow us to express the questions we want to ask about the slice of reality the model represents, and, furthermore, that the model provides answers to these questions. Referring again to Picasso's portrait of Gertrude Stein, if our question was about the color of her hair and Picasso had used green paint to

**Figure 1.9** *The Execution of Maximilian* by Edouard Manet (1867–1868).

portray it, then we would conclude that the portrait was not a good model at all—at least not for providing an answer to this particular question. On the other hand, if our interest were in knowing more about Gertrude Stein's personality and inner soul, and if Picasso's portait displayed these aspects of her in a particularly transparent manner, then the portrait would be thought of as a very good model indeed of its subject—even if he had painted Stein with green hair.

So the first and foremost test that a good model must pass is that it provide convincing answers to the questions we put to it. However, this is a kind of metacriterion that covers several more specific criteria for good modeling. Let's briefly deconstruct the somewhat vague notion of an answer into two of its most important component parts, in order to get a little better understanding of what this term might mean in a particular setting.

One thing an answer might mean is the ability to *predict* on the basis of the model what the real system will do next. For example, consider the Ptolemaic view of the solar system, in which the Earth sits at the center with the various planets moving about in orbits that are described by complicated epicycles. These curves are just the path traced out by a fixed point on, say, the rim of a coin as you roll it along the top of a flat table. Coins of different sizes give rise to different epicycles, and the Ptolemaic theory used combinations of these curves to characterize the planetary orbits. An excellent description it was, too, as piling more and more of these curves on top of each other gives increasingly accurate predictions about where the planets will be found next on their various tours about the Sun.

When it comes to providing an answer to *why* the planets move as they do, the Ptolemaic system is woefully inadequate, mostly because it is predicated on the wrongheaded notion that the Earth is the center of the solar system, rather than the Sun. As we have already discussed, the quintessential example of a model that does hold explanatory power is Darwin's famous Principle of Natural Selection, which is based on the survival traits of organisms. These traits offer an explanation for why certain species win out in the battle for survival in nature, whereas others fall by the wayside. The Principle of Natural Selection has certainly given deep insights into the process of life and death in the natural world, providing us with reasonably convincing explanations for many adaptations in species. For instance, it does a fine job of explaining why the peppered moth *Biston betularia* acquired dark pigmentation in industrial areas of Britain as an adaptive trait enhancing the moth's ability to avoid being seen by various predatory birds as it spread itself flat against the bark of trees in the area.

But a model like the Principle of Natural Selection, although of great utility for explaining what has already happened, is almost totally useless when it comes to telling us what will happen next. There is nothing in natural selection that would say that *Biston betularia* would develop the camouflage of darkly-pigmented wings to stave off extinction, as opposed, say, to developing the ability to burrow underground or secrete a poisonous fluid as an adaptation enhancing its chances of survival. So when it comes to prediction, the Principle of Natural Selection is as useless as the Ptolemaic epicycles are for explanation; they are complementary models, in the sense that one answers the question of prediction whereas the other addresses explanation (but, of course, in entirely different domains).

A better model would be one that could answer *both* questions of prediction and explanation. Newton's celebrated equations of particle motion serve as the standard example of a model of this kind. Through the relation $F = ma$ linking the accelerations $a$ of a collection of point particles to the forces $F$ impressed upon them, this model tells us not only where the particles will be at the next moment and with what velocities, but also gives an explanation of sorts (the external forces and the initial positions and velocities) for why they will be in that state.

Although answering questions is the *sine qua non* of a good model, there are other characteristics that tend to separate the good from the bad. Let me just list some of them in a more-or-less telegraphic fashion.

**Simplicity.**   One of the most revered principles of theoretical science is the doctrine "entities are not to be multiplied beyond necessity," attributed to the fourteenth-century English philosopher William of Ockham, and now known as *Ockham's Razor.* In everyday terms, this razor is a statement that an explanation should be as simple as possible, consistent with accounting for a given set of facts.

In modeling terms, Ockham's Razor can take many forms depending on the particular structure of the class of models available. For example, if we want to model the response of a passive electrical circuit, consisting of resistors, inductors, and capacitors, to given input voltages, then the simplest circuit might be the one containing the fewest number of these components. If we try to represent this problem mathematically by a set of differential equations expressing the currents and voltages in the circuit, then the condition on the elements translates into the mathematical requirement that there be as few differential equations in the system as possible, consistent with the requirement that the behavior of the mathematical model of the electrical circuit match the prescribed pattern. In semitechnical terms, this means that we want the dimension of the model to be as small as possible.

There are other notions of simplicity, as well, many of which we will encounter later in this volume, but almost all of them rely on providing a model that is in some sense the "most compact" of its type and that provides convincing replies to the questions we want to answer. So whether the criterion of simplicity is measured by the dimension of a system of differential equations, the length of a computer program, or the number of nodes in a directed graph, simplicity comes down to brevity of description.

**Clarity.** A good model is a clear model, one that can be understood and used to produce the same results by any interested investigator. In other words, nothing beyond the training necessary to understand the language the model is expressed in (for example, the symbolic language of mathematics) and the resources necessary to exercise the model (a supercomputer or a well-equipped chemical or medical laboratory, for instance) is needed to effectively use the model. Moreover, when the model is used properly it will provide every investigator with exactly the same answers; in short, it requires no divine inspiration or private interpretation to obtain answers from the model. A good model provides the same answers to all comers. This aspect of good modeling, incidentally, is one that most strongly separates the practice of theoretical science from that of many religions.

**Bias-free.** A good model tends to be objective. By this, I do not mean that it exists and produces its answers independently of human beings. Rather, a good model is objective in that it is independent of investigator bias, for example, the model expressing that the force of repulsion felt by two positively charged particles varies as the square of the distance between them. Thus, it is an inverse-square law, not an inverse-cubic, inverse-quartic, or inverse-$\sqrt{\pi}$ law. It remains an inverse-square law regardless of the investigator's political affiliations, the status of his or her bank account, or religious leanings. Note that this does *not* say that different investigators might not express the law in different terms. However, all these formulations must eventually turn out to be equivalent.

**Tractability.** As powerful computing capabilities become ever cheaper and more widespread, computer models are fast becoming the norm when it comes to picturing processes of both nature and humans. Indeed, that is the *raison d'etre* for this book. This suggests another criterion for good models, particularly for computer models rather than mathematical or experimental models: The model should be tractable. Roughly speaking, if the computational cost in obtaining answers to the questions we put to the model is greater than what we can afford, then we deem the model intractable. Computer scientists have captured this notion more precisely by talking of problems whose solution time grows exponentially with the size of the problem. But for us, all that is needed for a model to be bad is that the cost of computation be greater than what we feel able to bear. So, good models are affordable models in the currency of computation. Again, we will speak more precisely about this tractability issue many times as we make our way through the surrogate worlds presented here.

We have now looked briefly at several "fingerprints" that distinguish good models from bad, laying great emphasis on the point that good models provide convincing answers to the questions we put to them. After all, if models do not answer our questions, what is the point? A model that does not answer or at least suggest a question is useless, but how do we actually assess that answer?

For instance, suppose we have the computer model *FBPRO95*, and have asked it to tell us the result of Super Bowl XXIX. It replies with the answer shown on the scoreboard of Figure 1.2, namely, a 49er victory by a count of 21 to 7. This certainly constitutes a very definite prediction about what will happen. In fact, if I had kept careful track of each play of this simulation as it was run, that record would have formed an explicit play-by-play, turnover-by-turnover, touchdown-by-touchdown explanation of how this result was arrived at. Now for the $64 question: Is this prediction and explanation *reliable* enough for me to entrust my $64 to the Chargers with my friendly, neighborhood bookie? Depending on my attitude toward risk, as well as my knowledge of how well *FBPRO95* has done in predicting the outcomes of other games during the season, I come to a decision on the matter. As we know, the electronic gridiron inside my computer did not turn out to be such a good surrogate for the real one, at least insofar as this single instance of Super Bowl XXIX is concerned, as the 49ers soundly trounced the Chargers by 23 points in the actual game. Let's return to the art of modeling and take a little harder look at this question of just how much of our faith we can place in the answers that come to us from this kind of surrogate world.

## But Can We Trust It?

Both Plato and Aristotle held to a representational theory of art, in which artworks imitate real physical objects. But they differed radically on the matter of whether it was possible to gain either intellectual or practical knowledge about real-world things from their art-world representations.

In Plato's philosophy, true reality resides in the Eternal Forms or Ideas that make it possible to understand ordinary physical objects. Thus, according to Plato the way to gain knowledge is to directly encounter these "platonic" Forms. But since artworks are only imitations of physical objects, which are themselves only derivatives of the Forms, a work of art cannot provide us with any knowledge. As Plato described it in Book 10 of *The Republic:* "They [artworks] are at the third remove" from reality. So, for Plato art was not a source of knowledge or even of reliable opinion about objects of the real world.

Aristotle, although sharing Plato's view of art as presenting likenesses of things, argued that it is natural and beneficial for humans to learn by imitating and from carefully crafted imitations. In the *Poetics,* Aristotle noted that tragic poetry, unlike history, often expresses general truths, not just the facts of what actually took place. Rather, poetry tries to convey a feeling for what is likely to happen, generalizable truths about the sorts of things that probably or necessarily occur. But, he went on to say, it may be difficult to understand events that match these general truths, especially when these events are taking place in real time all around us.

As a result, Aristotle suggests that by composing an imitation of an action that is carried out on stage, the dramatist can display the same truth that is being shown by the real action, but in circumstances helpful to learning about the situation. The real action, possibly containing real death, tragedy and destruction, might distract us from the chance to learn.

But a suitably idealized imitation of the action may allow us to comprehend the principles that govern human activity. This is somewhat analogous to studying, say, the human heart or kidney. A plastic laboratory model of these organs might facilitate learning about their typical structure more effectively than dissecting the real heart or kidney of an anonymous corpse. In this case, the model reproduces and emphasizes the organ's essential structure and general features, but it eliminates the peculiarities and possibly repulsive and distracting aspects of a real organ.

Both Plato and Aristotle recognized that an essential aspect of art is that it is *different* from real things. Their views part company only on the point of whether we can learn about real things from this difference. For example, Plato would argue that there is nothing to be learned about late-nineteenth-century Parisian life from gazing upon Renoir's famous painting *Luncheon of the Boating Party*. Aristotle, on the other hand, may well argue that this painting encapsulates an enormous amount of information about how people of a certain social class interacted and how they lived in *fin-de-siècle* Paris. Nevertheless, the portrait of Parisian life shown by Renoir is certainly not the real thing, and to believe it is would be like having a member of the audience jump up and call for the police during the scene in Shakespeare's play *Othello* when Othello strangles Desdemona on stage. In both the play and the painting, a crucial aspect of understanding the artwork lies in realizing that art objects must be different from real things.

Interestingly enough, postmodern artists try to reduce the distance between art and real things. As an illustration, consider the artist Robert Indiana who paints pictures of bull's-eye targets that are at the same time real targets and imitations of real targets. Now suppose you hung a real target next to such a painting. Would it be acceptable for an archer to shoot arrows at the Indiana painting? Or would an art afcionado object that you should restrict your shooting only to the target, even though the target and the painting look exactly alike? Does the Robert Indiana painting tell us anything about real targets by imitating them in paint on a canvas? That is, do we learn anything about the real system from a model that is indistinguishable from it? Hard questions.

As we consider various computer models of the real world in this book, we will continually encounter questions like these relating to how the model teaches us something about real things. Sometimes, the model

will simply try to duplicate as closely as possible the real-world phenomena, as with the football simulation *FBPRO95* already considered.
In other cases, the model will be a deliberate caricature of the real system, specifically designed to exaggerate some aspect of the system at
the expense of fidelity in other areas. Such an exaggeration aims at
drawing our attention to a particular feature of the real-world system
in much the same way that political cartoonists used to draw distorted
version's of Richard Nixon's ski-jump nose as a quintessential feature of
his physiognomy.

In contrast to Plato's view, I believe that there is much to be learned
from both types of models. It all depends on the nature of the questions
that we want answered, and upon how much we actually know about
the underlying laws governing the ways the various parts of the system
under study interact with each other. In football and road traffic, we
know a lot about the parts and their interactions; in cellular biology, on
the other hand, we know only a little. As with all computer exercises, it's
garbage-in, garbage-out, and the faith we place in the model's answers is
inversely proportional to the amount of garbage that goes in. The reader
can assess the level of garbage-in as we make our way through the models
sprinkled throughout the remainder of this book. Now let us consider
briefly the principal reasons for building such computational models.

## "Hypotheticality"

Some years back, Professor Wolf Häefele, director of the German Nuclear Research Center in Jülich, coined the term *hypotheticality* to refer
to situations in which we have to make life-or-death decisions but are
unable to perform the experiments or tests needed to gain much-needed
information about the situation before having to make the decision. At
the time, Häefele's concern was mostly with the possible dangers—real
and fanciful—seen by environmentalists and other citizens concerned
with the widespread use of nuclear power. But the term has much broader
currency than merely as way of describing the fact that there is no way to
confirm or deny a particular claim about hypothetical dangers of nuclear
reactors. It can refer equally well to almost any assertion one cares to
make about the behavior of what we now call a *complex system*. Let's
look at a few examples.

## Global Warming

Suppose the amount of carbon dioxide dumped into the atmosphere is doubled from its current levels. What effect will this have on the average global temperature 50 years from now? This is a standard scenario used by climatologists to characterize the problem of global warming. The answer to the question is anyone's guess, with different groups of investigators coming up with estimates ranging from no effect at all to an increase of up to five degrees Celsius, the so-called greenhouse effect. No one really knows, but almost everyone cares, because a major temperature increase can have dramatic effects on everything from agricultural yields to the continued existence of the Antarctic ice cap.

## Genetic Engineering

A few years ago, amidst considerable concern and consternation among environmentalists, permission was granted to a an agronomist in Nevada to plant a strain of tomatoes that had been genetically engineered to have a better flavor at the end of its extended shelf life at the supermarket. Since that time, more than 60 plant species have been genetically engineered to provide plants that would boost food production by reducing losses caused by pests and diseases, while requiring far less pesticides.

Despite these unquestioned successes, the scenario now unfolding is far less rosy than that envisioned in the heady days of the late 1970s when genetic engineering first sprang forth from the cozy confines of the world's biomedical research laboratories. At that time, it was felt that plant breeders would no longer be limited to working with already existing traits in a plant's genome. Rather, desirable traits from a completely unrelated organism could, in theory, be transferred to a plant by direct insertion of the gene coding for whatever traits one wanted. But, as just noted, in two decades of trying, genetic engineers have managed to influence only a small number of plants in ways that actually help increase food production. What happened?

Basically, the reason why genetic engineering has been at best an evolution rather than a revolution is a combination of science and money. It turns out that the traits that influence a plant are a lot more subtle than genetic engineers had counted on, much in the same way that the features characterizing human intelligence are a lot more difficult to pin down than early researchers in artificial intelligence had suspected. For instance, companies that have developed bug-resistant seeds now tell farmers to plant only about 80 percent of their land with these seeds, so that there will be enough susceptible bugs left over each year to reproduce most of the population. In this way the seeds should remain effective for 50 years or so, rather than just a decade or less. Moreover, from a business standpoint, there just are not very many crops that a profit-oriented firm can afford to do much research on. And since private industry funds over 75 percent of genetic engineering research on plants, only a very few high-value crops—cotton, tobacco, maize, potatoes, tomatoes—have received the attention of genetic engineers. Not at all the kind of revolution that people were speaking of 20 years ago, when dreamers, schemers, and scientists were all rhapsodizing over a genetic solution to the problem of world hunger.

## Monetary Policy

Shortly after the end of the Gulf War, the American stock market took off on a trajectory that sent the market averages soaring to all-time highs. The reason for this meteoric rise is generally thought to be the historically low interest rates during the early 1990s, which not only led to higher corporate profits but also left investors with very few attractive alternatives to stocks in which to park their money. Following this line of reasoning, conventional wisdom would argue that when the Federal Reserve began ratcheting-up interest rates in early 1994, the U.S. dollar would rise against other currencies and the market averages would begin to tumble. But not so. In fact, the dollar continued its lethargic performance against the German mark and the Japanese yen, while the Dow Jones Industrial Average actually rose 80 points during the course of 1994, in the face of six increases in interest rates by the Fed. So much the worse for conventional wisdom, and so much the better for hypotheticality!

## AIDS Vaccine

In early 1995, a furor broke out over a plan by the government of Thailand to inject people with an experimental AIDS vaccine developed in the United States. The heart of the controversy was the fact that the vaccine was based on the subtype of the HIV virus most common in the United States. But there were two genetic subtypes circulating among the population of drug users in Bangkok who were to form the subjects for the trial. Thus, one group of scientists claimed that the vaccine was a mismatch for at least one, if not both, of these Thai subtypes, and so its chances of doing any good were vanishingly small.

Other scientists, including a number of Thai researchers, said there was no hard evidence to support the claims that the vaccine would fail to protect against the Thai subtypes, and that until a trial was done, no one would ever know whether the vaccine would be effective. Furthermore, their argument continued, if nothing were to be done, the AIDS epidemic in Thailand would escalate. Again, hypotheticality.

What all these examples have in common—and the list could be greatly extended in almost any direction one cares to look—is that we are faced with having to make life-and-death choices, literally, about a system whose workings are almost a total mystery to us. Each of the processes sketched above is an example of a complex system, one consisting of a large number of individual agents—investors, virus molecules, genes—that can change their behavior on the basis of information they receive about what the other agents in the system are doing. Moreover, the interaction of these agents then produces patterns of behavior for the overall system that cannot be understood or even predicted on the basis of knowledge about the individuals alone. Rather, these emergent patterns are a joint property of the agents *and* their interactions—both with each other and with their ambient environment. The ability of such systems to resist analysis by the traditional reductionistic tools of science has given rise to what is now called the sciences of complexity, involving the search for new theoretical frameworks and methodological tools for understanding these *complex systems.* This volume tells the story of what is perhaps the largest cannon in this arsenal, microworlds constructed inside the digital computer.

## From Real Space to Cyberspace

Sometime in the sixteenth century, Galileo ushered in the practice of modern science by introducing the idea of a controlled, repeatable laboratory experiment. As we know from the time-honored scientific method, a crucial part of the construction of a decent scientific theory of anything is to be able to do such experiments in order to test hypotheses about the phenomena under investigation.

The problem with complex systems like nuclear power plants, the AIDS virus, the economy or the human genome is that we cannot do the kinds of experiments needed to create a reliable model and theory of their operation. For example, it doesn't take too much imagination to envision the reaction of the Securities and Exchange Commission, not to mention the global financial community, if someone were to propose moving interest rates up, say, 500 basis points for a couple of weeks in order to test some new theory of currency and stock-price fluctuations. It's just not possible to do this kind of experiment because the operation of the system is too central to daily life to mess about with it in this way. Sometimes, as with the nuclear reactor, experiments aimed at understanding the limits beyond which a meltdown will occur also cannot be performed, simply because they're too dangerous; the experiment could easily kill the experimenter. And so it goes with the other systems considered.

So it goes—until now. For the first time in history we are in a position to do bona fide laboratory experiments on these kinds of complex systems. No longer do we have to live in the shadow world of hypotheticality, or just break off bits and pieces of the actual system and study these fragments in isolation with the hope that we can then re-assemble these chunks of partial knowledge into an understanding of the overall system itself. But now, thanks to the availability of affordable, high-quality computing capabilities, we can actually construct silicon surrogates for these complex, real-world processes. We can use these surrogates as laboratories for carrying out the experiments needed to be able to construct viable theories of complex physical, social, biological, and behavioral processes. In many ways this leaves us in the same position that physicists were in at the time of Galileo. We now have an essential tool that can be used to create theories of complex systems, theories that will ultimately compare favorably with the theories of mechanical processes that Newton and his successors developed to describe simple particle systems. The remaining chapters of this book describe some of the halting first steps on the way to such theories.

By now, it has probably not escaped your attention that it is but a small step—conceptully, if not technologically—from the passive observation of "simworlds" to the experiencing of these worlds as an active participant. This, of course, is the theme song of the *virtual reality* movement, which aims to create silicon universes that are indistinguishable to our senses from the real thing—or at least what might be some kind of real thing. Although the idea of donning a video helmet and a set of data gloves to enter an electronic spitting image of the National Football League seems farfetched to many, virtual reality afcionados would regard this as the mildest among the many possibilities that technology will admit during the lifetime of most of us. In fact, one could make a good argument for this sort of active participation already being available in, for example, the flight simulators in which both commercial and military pilots hone their skills.

Our accounts in this book will not delve much into such scenarios for active participation, because not only has the popular literature on avant garde computing already scouted out this territory in great detail, but the models and the technology for actually jumping into electronic worlds is still not in a state to be taken very seriously—with a few honorable exceptions, such as flight simulators. But as we move along from artificial world to artificial world in the chapters that follow, you bear in mind the possibility of actually visiting and even taking up part-time residence in these worlds sometime in the not-so-distant future. The days of being imprisoned within the messy, often unpleasant realities of life on Planet Earth are almost at an end. On this dynamic, uplifting note, let us get on with our account of electronic worlds, both near and far.

# CHAPTER

# 2

# Pictures as Programs

## That's Life?

In the summer of 1975, Chris Langton almost killed himself in a hang-gliding accident on Grandfather Mountain in North Carolina. But *almost* only counts in horseshoes, and this near fatality led directly to the exploding field Langton later christened *artificial life.* During the long months of convalescence, Langton devoured piles of books on biology, philosophy, computing, genetics, mathematics, even some science fiction. The end result of this reading was a growing conviction that there was nothing about living organisms that could not be recreated within the cozy confines of a computing machine. By the time he was released from the hospital, Langton's life's work was determined: to create nothing less than "life in silico," as opposed to life *in vivo.*

Following a long and tortuous path leading from Tucson to Ann Arbor and finally to Los Alamos and Santa Fe, Langton persevered in his quixotic quest. By the mid-1980s he was ready to call for a show of

37

hands. Was his search for the essence of life inside a machine simply an eccentric dream of a few fanatics like himself? Or was there really a growing community of like-minded souls scattered across the academic landscape ready to do battle with the biologists about what it means to be alive? To smoke out the committed believers that Langton felt sure were out there, he convinced his then employer, the Los Alamos National Laboratory, to cosponsor, along with the Santa Fe Institute and the Apple Computer Corporation, the first international workshop on the synthesis and simulation of living systems.

By all accounts this gathering was an unqualified success. One of the most popular programs presented at the inaugural meeting was something cooked up by noted biologist Richard Dawkins of Oxford University. What Dawkins showed the gathering was how the process of genetic mutation coupled with natural selection could be used to "breed" objects that he christened *biomorphs.*

In Biomorphland, one begins with a kind of stick figure. This skeletal object is then mutated in accordance with a set of rules, creating a spectrum of offspring that are each determined by a single genetic mutation. The human investigator then plays the role of nature, selecting (on aesthetic grounds or otherwise) one of these objects to move on to the next generation, whereupon the process of mutation and selection is repeated. Figure 2.1 shows one of Dawkins's evolutionary experiments, and Figure 2.2 depicts the end result of several such histories.

The biomorphs created in Dawkins's computer are truly alien life-forms—but ones bearing an eerie resemblance to various sorts of insectoid forms we see in many earthly organisms. Yet all of Dawkins's creatures emerged from a few simple rules applied repeatedly to an ancestral form, helped along by a bit of godlike investigator interference from time to time to select those creatures that should die and those that should move on to the next generation.

A few years after Dawkins's work, an even simpler procedure for creating biomorphs was cooked up by Clifford Pickover. The key to his entry into Biomorphland is to employ the same scheme used to create many of the objects seen in the mathematical theory of fractals. Let me briefly sketch the general idea.

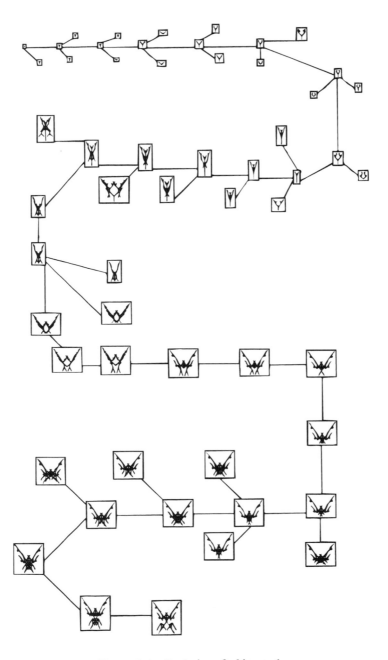

**Figure 2.1** Evolution of a biomorph.

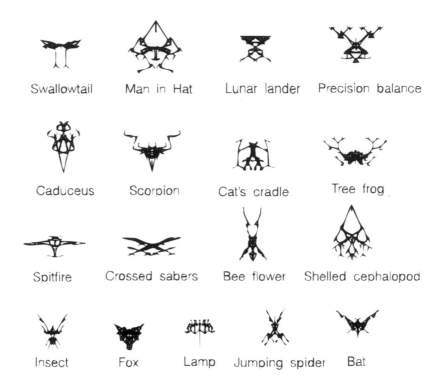

**Figure 2.2**  Some residents of Biomorphland.

We begin by picking an initial point in the $x$-$y$ plane. We then create the next point by the following rule: If the current point lies at the point $(x, y)$ in the plane, the next point has coordinates

$$x \to x^3 - 3xy^2 + \frac{1}{2},$$
$$y \to 3x^2y - y^3.$$

This rule is then applied to the new point and so on, thereby generating a sequence of points. This way of creating points by repeated application of a fixed rule is called an *iterative,* or *generative,* process, with the rule being the generator and the initial point often called the *seed.* If after a large number of iterations, say a few thousand, the current point is less than some prescribed distance, say, 10, from the origin, then we color the starting point of the sequence black; otherwise, we leave it white. We then pick a different starting point in the plane and repeat the process.

**Figure 2.3**   A radiolarian biomorph.

Choosing many different starting points and using this coloring scheme, we end up with the biomorph depicted in Figure 2.3. What immediately catches the eye about this biomorph is its uncanny structural resemblance to a real-life organism called *radiolaria,* which includes things like the amoeba. These are single-celled creatures with an intricate symmetrical exoskeleton that often has many spines extending outward. Figure 2.4 shows some of the many different types of biomorphs that can be generated using this same procedure, simply altering slightly the specific rule for generating the points. (*Note:* For the sake of brevity, the generative rules shown in the figure are expressed using the notation of complex numbers.)

The striking aspect of these two sets of computer experiments is not that the simple rules employed by Dawkins and Pickover are in any way similar to what nature uses to create living things. Rather, it is that such seemingly complicated geometrical shapes and forms can arise, from such humble beginnings, at all! In both cases the complexity of the final forms seems to be enormously greater than the complexity of the rules giving rise to them. Just to hammer home this point, let's look at one more example of this.

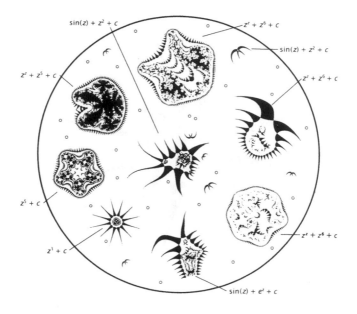

**Figure 2.4**   Different biomorphs generated by iteration.

## Artificial Plants

In the mid-1960s, Dutch biologist Aristid Lindenmayer began developing a scheme by which to describe the growth of plants by following a set of rules similar to those presented above. Lindenmayer's scheme, now called an *L-system* in his honor, can be viewed as a series of rules by which the various parts of a plant can be created. Here is a very simple illustration of the underlying idea.

The blue-green algae *Anabaena catenula* is composed of two types of cells strung out on a filament. Call these types $a$ and $b$ to distinguish the differences in the size and readiness to divide of the cells making up this plant. Under a microscope, a filament of this plant appears as a sequence of cylinders of various lengths, with the $a$ types being longer than those of type $b$. Moreover, each cellular type has a label, $l$ or $r$, specifying the positions (left or right) where daughter cells of types $a$ and $b$ will be produced. So, taking these labels into account, there are a total of four different cellular types. Let us call these types $a_r$, $a_l$, $b_r$ and $b_l$. Suppose we begin with a single cell of type $a_r$. To create new parts

42

(cells) of *Anabaena* from this ancestor cell, we employ the following four rules:

$$\text{I: } a_r \rightarrow a_l b_r, \quad \text{II: } a_l \rightarrow b_l a_r, \quad \text{III: } b_r \rightarrow a_r, \quad \text{IV: } b_l \rightarrow a_l.$$

(The first rule reads: Replace a cell of type $a_r$ by the two adjacent cells, the new cell on the left being type $a_l$, and the one on the right is of type $b_r$. I will leave it to you to give a similar verbal description for rules II–IV.)

This set of four rules constitutes what is called a *rewrite system,* because it tells us how to rewrite any of the four cell types to a new set of cells at each step of the developmental process. Figure 2.5 shows the result of applying these rules one at a time from a single initial cell of type $a_r$. Starting from the single cell of type $a_r$ at the top, we obtain the second row of the figure by using rule I. We then apply rule II to the cells on this new row, and so on, ending up with the cells at the bottom row after four steps. Note, however, that it is not necessary to use the rules in sequential order I, II, III, and IV, as in this example. We could just as easily have applied the same rule several times in succession or mixed the rules in other ways, thereby creating different types of plants.

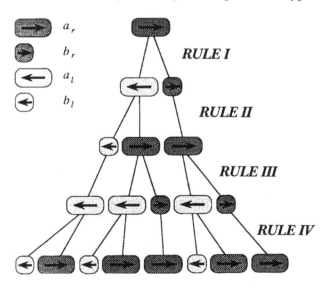

**Figure 2.5**   Four steps in the development of the filamentous plant *Anabaena.*

Although definitely a plant, *Anabaena* is about as primitive as plants get. Moreover, the version of this plant shown in Figure 2.5 is something akin to a cardboard cutout version of the real thing, being simply a two-dimensional model showing only the most basic aspects of the growth of this blue-green algae. But the basic idea of growing a plant from a set of simple rules like those above has much broader currency, and has been pushed to dramatic extremes by Lindenmayer's coworkers, especially P. Prusinkiewicz. Although there is neither room nor need to go into the details here, Figure 2.6 shows just how realistic plants generated by a set of L-system rules can look. The figure displays the developmental path of an electronic hawkseed flower obtained by using L-system rules only slightly more complicated than those for *Anabaena* (along with some more sophisticated computer graphics procedures to render the result in living color).

**Figure 2.6** Development of a hawkseed flower by following L-system rules (see Plate III). P. Prusinkiewicz, Mark Hammel, and Eric Mjoiness, "Development of Hieracium umbellatum" in *Proceedings of SIGGRAPH '93*. ⓒ Copyright 1993 Association for Computing Machinery. Reprinted by permission.

## Models or Reality?

The biomorphs created by the simple rules of Dawkins, Pickover, and Lindenmayer are certainly not even close to mimicking the real-life organisms we see crawling around on the surface of the Earth today. For example, real organisms are three-dimensional objects with a richly detailed internal structure; biomorphs are two-dimensional creatures, having no volume at all. Moreover, biomorphs do not actually *do* anything. They are simply mathematical objects whose geometrical form happens to bear a striking resemblance to the forms of some types of living things. Is there any way to defend the claim that the computational rules, or programs, leading to the biomorphs serve as "models" for the form, if not the dynamics, of real-world organisms? Or are they just computational curiosities, having no more relevance to biological organisms than, say, a model car has to a fire-breathing Ferrari?

Evaluating these computer models for biomorphs against some of the yardsticks considered in the opening chapter—simplicity, clarity, bias-free, tractability—the models stand up pretty well. After all, hardly any rule could be much clearer or simpler than the rule given earlier for generating Pickover's radiolarian. Nevertheless, these computer models all fail pretty badly when it comes to their "believability." It is just very difficult to take seriously the idea that a simple cubic formula like that for our silicon versions of radiolaria is even close to the rule that Mother Nature uses to create the real-world radiolaria—if she uses any rule at all! Being able to create a form that mimics a real radiolarian and being able to create the radiolarian itself are simply two very different things. Therefore, although these computer programs may suggest that the natural processes of biological development are rule-governed, we probably would not want to bet the monthly mortgage payment on their being even close to whatever actual rule(s) cells follow in forming the structure of a real, live radiolaria, *Anabaena,* or anything else. But there is another way to look at these experiments that makes these look like very good models indeed. In fact, they look like the best of all possible models—the real thing itself.

We have been speaking of models as being in some way representations in symbols of natural or human phenomena like living organisms, football games or the solar system. The purpose of such models is to enable us to predict and/or explain something about the real-world process they represent. In short, the standard of judgment as to whether the model is good or bad is grounded in how well the model answers our questions about the world of people, places, and things. But what if we break this connection with the real world? What if the world of objects and processes to which the model refers is itself a symbolic one inside our computing machines? In that case, what we have called the model itself becomes the system. Let us see how.

Consider once again the rule for generating the radiolarian shown earlier in Figure 2.3. In the terms of chapter 1, this is a computational model. From the standpoint of telling us about the shape of such an organism, the model is fairly good, although it fails pretty miserably when it comes to answering just about any other kind of question about the structure or behavior of a radiolarian. But let us drop this association with the real world and consider a world in silicon whose inhabitants are the critters created by Dawkins and Pickover shown earlier in Figures 2.2 and 2.4. In this case, the rule ceases to be a *model* of a radiolarian and instead becomes the very *definition* of the biomorph shown in Figure 2.3. In that artificial world, the rule for generating the radiolarian *is* the creature; hence, this rule is the best possible model of the real thing, namely, the thing itself.

This latter line of argument may seem strange to those accustomed by history and habit to think of models as always referring back to the familiar world of carbon-based organisms, but that is simply a prejudice. There is no reason at all to think that our everyday world has any privileged ontological status and is any more real than the worlds we can create *in silico* rather than *in vitro*. These computer worlds are just as real as our own real world, provided they are looked at from *inside* the computer rather than from the usual vantage point on the outside. This internal/external dichotomy is a theme that will regularly pop up during our deliberations in this volume, so there is no need to belabor it here. The many examples of such so-called artificial worlds that we will see in subsequent chapters will make the point much more forcefully than a ton of vague generalities and armchair philosophies. Keep in mind that the models we shall encounter must always be judged against the particular world they refer to, and that our computers have finally opened up the possibility to create an infinite set of alternatives to the one real world we have been restricted to in the past.

Nowadays, if you want to examine the properties of a world in which gravity is negative or stock dividends differ from their Wall Street values, you can simply build that world in your computer and see what happens, but these worlds are still subject to the constraints imposed by the workings of the real-world computers serving as their hosts, so to speak. With these ideas of would-be worlds on the table, let us take a short breather and look at the way computing machines actually work in order to whip up these worlds for all occasions.

## The Genesis Machine

The year 1935 was a good one for computing. In fact, it was the year that Alan Turing set out to solve a famous unsolved problem in mathematics, Hilbert's Decision Problem, and along the way to a negative solution ended up inventing the theoretical foundations for what we now call the theory of computation. The crown jewel in Turing's work was a kind of theoretical computer that we now call a *Turing machine.* In rough outline, what this gadget consists of is an infinitely long tape ruled off into squares, each of which can contain the symbol 0 or 1, together with a scanning head capable of being in any of *n* configurations, or states. These states can be thought of as the possible positions of a pointer inside the scanning head. The head can move along the tape, square by square, performing one of the following actions at each step:

PRINT 1 ON THE CURRENT SQUARE.
PRINT 0 ON THE CURRENT SQUARE.
MOVE LEFT ONE SQUARE.
MOVE RIGHT ONE SQUARE.
CHANGE THE CURRENT STATE TO ANOTHER STATE.
RETAIN THE CURRENT STATE.
HALT.

So that's it. Just these seven possibilities at each step. Yet Turing showed that a machine displaying this seemingly very limited repertoire of actions is capable of computing anything that can be computed. In fact, he proved that there exists a "universal" version of this computer, a so-called *universal Turing machine,* that is capable of emulating the action of *any* computing machine whatsoever. Thus, whatever your Macintosh, SPARC workstation, or Connection Machine 5 can compute, so can the UTM (but, alas, much more slowly, because computing speed is hardware-dependent). It is not especially important for our purposes here exactly how the UTM operates, so the interested reader can refer to the references for the details. What definitely is important for us, though, is how such a machine can be used to code anything about the world that can be expressed in symbols—linguistic or otherwise.

Everything that can be represented by a UTM or any other computer must somehow be expressed as a pattern of 0s and 1s in the memory of the machine. This means that if we want to represent words or pictures in such a machine, they have to be reduced to such patterns. For the sake of definiteness, let us suppose our interest is in expressing statements in English. Because expressions in English are composed of letters, punctuation marks, and spaces, we need a convention for representing the characters of the English alphabet, along with these other symbols, as strings of 0s and 1s. The convention that is almost universally employed nowadays is something called the American Standard Code for Information Interchange (ASCII), which assigns a unique pattern of eight 0s and 1s to every symbol used in written English. For example, the ASCII code for some characters are

$$A = 01000001 \qquad I = 01001001 \qquad M = 01001101$$
$$! = 00100001 \qquad ␣ \text{ (space)} = 00100000 \qquad ? = 00111111$$

Using this ASCII coding scheme, we can express the interrogative sentence "AM␣I?" in ASCII as the binary string

$$01000001|01001101|00100000|01001001|00111111,$$

while the declarative sentence "I␣AM!" is coded as

$$01001001|00100000|01000001|01001101|00100001.$$

A coding scheme like this may seem more than just a little bit tedious for humans to use in expressing even a short statement involving just a few sentences, let alone a substantive block of text like *War and Peace*, but it's no problem at all for computers to throw such strings around with abandon—as well as accuracy. Just in case you're wondering, the representation of pictures is almost as simple as representing text. Basically, one divides the picture into a large grid of picture elements, or pixels, assigning a particular color value to each of these individual "atoms" forming the overall graphical pattern. There are a variety of schemes that people use to code these colors, schemes that work in a fashion completely analogously to the way the ASCII procedure works to code the characters of written English.

By using these codes, what we end up with is a way to express any written or visual object as a binary string. In effect, these coding schemes enable us to represent the world symbolically inside a machine. But that's not all. Once we have represented a particular idea or pattern by some long binary string, a program can be written to transform that string into another such pattern of 0s and 1s. This new string, in turn, can then be *decoded* into some statement in words and/or picture elements. In this way statements can be presented to a computing machine as inputs, and new statements received from the computer as outputs. If this input/output process continues for several rounds, we have what one might term a *dialogue* with the machine. At exactly this point we find much of the debate surrounding "thinking machines," or what's more commonly called *artificial intelligence* (AI).

Proponents of AI argue that it's possible for programs to be either designed or evolved that will make such dialogues indistinguishable from those we might have with another human being. Anti-AI types say, No way! The one thing that both sides agree on, though, is that at present the kind of programs we have written are nowhere near to being able to fool a human interrogator into thinking the machine is actually thinking like you and me. Nevertheless, several programs have done a rather good job of imitating a human, at least in dialogues restricted to certain domains of discourse. Just to get a feel for what's possible in this regard, let's look at one such dialogue with a program called *Racter,* a shortening of the term *raconteur.* What follows is a snippet of dialogue between Racter (**R**) and an artist, Joan Hall (**J**).

**R:** Would you please speedily yet cleverly tell me your name?
**J:** Joan Hall.
**R:** Shall we hastily or leisurely talk a little?
**J:** No.
**R:** Now, Hall, aside from being loving, what are you, and from where?
**J:** An artist from New York.
**R:** Joan, in view of the disgusting fact that you're an artist, listen to this. Years and days are spooky, Joan, now that my dissertation is insane.

Some cosmologists from New York were singing to some other cosmologists. "Anyway our commitments are insane." Briskly the first cosmologist sang. By the same token the ambiguity is important, at all events, Joan, such tales are captivating. Do you not agree?

**J:** I suppose so.

**R:** You cried "I suppose so." Anyhow captivating cosmologists will never be arrogant. "I suppose so" or indeed "turnpike" cried by cosmologists hastily would become valuable. Suddenly they prepare for a commitment; yet enthralling hours are as minutes. In a millisecond we will get ready for enrapturing and arrogant dissertations. Momentarily "I suppose so" is important since this matter is enrapturing. You are exhausted, nevertheless an artist will speculate about neutrinos. Are these your views?

**J:** My view is that you are out of your mind.

**R:** Now wait. You sang "I suppose so," you now yodel "my view is that you are out of your mind"; drunkenly and cunningly I have now glimpsed your angry anguish. You, an artist, yodel "my view is that you are out of your mind."

OK, I think you get the idea. A conversation with *Racter* is eerily similar to talking with a schizophrenic. *Racter*'s ramblings seem to constitute a coherent unit of conversation, but one that wanders aimlessly from topic to topic with no overall organizing center or any apparent goal. However, it is difficult to pinpoint where, exactly, *Racter*'s replies differ from those of a highly excitable patient in a psychiatric hospital. All this from a simple program relying upon nothing more than an endless sequence of iterative applications of a few simple rules not unlike those used above in generating the biomorphs. The only real difference is that now the morphological objects are words and sentences rather than biological forms. While it would take us too far afield to go into the fine details of how *Racter* works, it's rather easy to describe its general structure.

*Racter* begins by selecting an item at random from a file. If the item is a literal statement, like "Tell me more," *Racter* simply prints it out directly. But it's far more probable that *Racter*'s random choice will yield a command, which will send the program off into other files that may themselves contain commands. When the initial command has finally been executed, *Racter* goes back to its files and randomly selects another element, thereby beginning a new cycle.

In order to begin a sentence, *Racter* must choose a form for the sentence, either at random or by some other rule that takes into account the program's recent dialogue. And in order to give its sentences some punch, *Racter* uses various identifiers, which are two-letter labels attached to words that are associated with different words. These identifiers cause *Racter* to make associations between successive words and sentences. As an example, *Racter* might come up with a sentence template like

THE noun.an verb.3p.et THE noun.fd.

where *an* is an identifier for *animals,* while *et* is associated with *eating* and *fd* is a tag for *food*. This template constitutes a rule for sentence generation. The program first searches its files for a noun, but only those nouns bearing the *an* tag. Thus, it would choose at random among nouns like *lion, tiger,* and *cat*. Next, having selected an animal, for instance *tiger, Racter* chooses a random verb bearing the identifier *et*. These verbs involve eating and might include such words as *eat, chew,* and *chomp*. Again, a random choice from among this set might lead to *chew*. Having chosen the verb, the program next forms the third-person past tense indicated by the code *3p* in the sentence template structure. Finally, the directions tell *Racter* to pick a noun carrying the food tag, *fd*. This may lead to the word *trout*. Putting all these choices together, the program spits out the sentence

THE TIGER CHEWED THE TROUT.

Before leaving *Racter,* it's worth noting that the program can do much more than just select words with identifiers. It can generate its own sentential forms and even its own command strings, inserting them into the conversation at will. The interested reader can find out more about *Racter* by consulting the material cited in the references for this chapter.

The examples of generating biomorphs and schizophrenic dialogues shows the power computing machines have to fabricate new kinds of structures that have never existed before, a kind of creativity if you like. This sort of mechanical creativity lies at the heart of the issues we want to address in this book, and they will be explored in more detail in areas a lot closer to human concerns like financial markets and road-traffic networks later in the chapter. However, creating linguistic and artistic forms is most definitely not the way most people either use or even think about using computers. Rather, the commonly held view of such machines is that they are glorified calculators whose job is to carry out numerical computations in a variety of settings, from keeping track of bank accounts and phone bills to making sure that ticket reservations match the seats available at the ballpark or at the theater. Just for the sake of comparison and contrast with our use of the computer as a symbol processor, let us look in a bit more detail at how the computer works as a calculator. This will aid in understanding the difference between how computers are usually used in science as machines for doing arithmetic computations, and how this contrasts with our use of them here as devices within which to construct new worlds.

## The Electronic Abacus

Certainly among the most important and revered theorems in number theory are Euclid's proof of the infinitude of primes (numbers divisible only by 1 and themselves) and the Fundamental Theorem of Arithmetic, which states that if a number is composite (not a prime) then it can be decomposed (uniquely) into a product of prime factors. For example, the number 168 is clearly not a prime because it is divisible by 2. Moreover, it is an easy pencil-and-paper exercise playing around with a few small primes to see that $168 = 2 \times 2 \times 2 \times 3 \times 7$. This is the only way to write this number as the product of primes. However, it is *not* an easy exercise to see that the number 512,461 is composite, having the factors 31, 61, and 271. When it comes to a number like

$$2^{59} - 1 = 179,951 \times 3,203,431,780,337,$$

which was discovered to be composite more than a century ago, techniques far beyond trial-and-error hand computations are required.

With the advent of the digital computer, the search for schemes to determine rapidly if a number is composite and, if so, to find its prime factors has become something of a cottage industry, and not just out of academic curiosity, either. In the late 1970s, the first type of coding scheme appeared by which one could easily code a message into a large composite number, and even publicly announce the way anyone wishing to send you a message should encode it. Decoding the message, though, required either knowing the prime factors by which the message was encoded into the composite number or somehow finding those prime factors by direct calculation. Because creating uncrackable codes is a crucial component of both the military and business intelligence business, government agencies around the world—especially the U.S. National Security Agency—began taking a very hard look indeed at methods for decomposing composite numbers into their prime components.

The problem of prime factorization represents a quintessential example of how computers are used as unimaginably large and powerful calculators to carry out computations that would take eons if done with paper and pencil in the old-fashioned way. The essence of all known schemes for factoring numbers involves a volume of calculation that boggles the human mind, with even the largest supercomputers still unable to factor numbers greater than about 150 digits or so in a time less than that of the lifetime of the universe.

Another well-chronicled problem in which the computer-as-calculator played the decisive role is the famous Four-Color Map Problem. This puzzle asks if it is possible to color any map that can be drawn on a piece of paper (more precisely, on the surface of a sphere) with no more than four colors. Almost from the time the problem was first stated in 1852, it was clear that four colors were necessary as shown by the maps in Figure 2.7. The one on the left actually uses five colors, while three colors is not enough to color the one on the right. But to show that there could be no map requiring more than four colors defied the efforts of the world's greatest mathematicians for more than a century.

But in 1976, Kenneth Appel and Wolfgang Haken startled the mathematical world in two very different ways. First, they offered a proof that four colors does indeed suffice to color every planar map. But of more lasting importance is that the method of proof they used caused us to

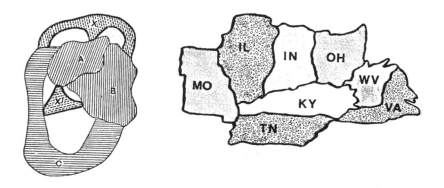

**Figure 2.7** Two maps each requiring four colors.

rethink completely the notion of what we mean by a mathematical proof. This was because the procedure employed by Appel and Haken to resolve the Four-Color Conjecture involved checking 1,936 special configurations, where the check of each configuration took many hours of supercomputer time. For the first time in history mathematicians were faced with a proof that could not be done by hand, and to make matters even fuzzier, it is known that the computer employed by Appel and Haken makes about one numerical error per thousand hours of operation. It would not be unreasonable to expect an error or two in the computation. This leads us to ask if a proof is really a proof if no one can check it and it contains several errors.

In the 20 years or so since the original work by Appel and Haken, the mathematical community has come to accept their computer proof of the Four-Color Conjecture, mostly because the approach is sound and the same results have been obtained using other computers employing different kinds of hardware and different programs. Nevertheless, it remains an open issue in the philosophy of mathematics whether the use of a computer to exhaustively check a large—but finite—number of cases of anything really constitutes an honest-to-god mathematical proof.

Both the problem of prime factorization of composite numbers and the computer solution of the Four-Color Conjecture are examples of how computers are used as overgrown calculators, to do numerical computations that humans could do, in principle, but that computers can do a lot faster. Just to show that the computer-as-calculator is not just a new tool for mathematicians, let's look at a couple of problems from biology and physics whose solutions involve using the computer in exactly this same fashion.

### Protein Folding

The fundamental building blocks out of which all living things are made are proteins, each of which can be thought of as a beaded necklace consisting of hundreds or even thousands of colored beads, each bead being one of 20 colors, or types, called amino acids. These amino acids are strung together in accordance with the instructions contained in the cellular genetic material, the DNA, the resulting protein chain of amino acids then folding itself into a particular three-dimensional configuration that determines how it will function in the organism. An example of this

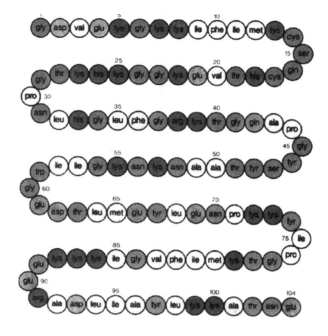

**Figure 2.8**   The one-dimensional structure of *cytochrome c.*

process is shown in Figure 2.8 for *cytochrome c,* a protein found in the respiratory tract. It consists of a string of 104 amino acids labeled by their short names, whose one-dimensional structure is shown in the figure. Figure 2.9 shows how this string of beads then folds up into its final, three-dimensional working configuration.

This final folded structure seems to be essentially one of thermo-dynamic stability, in which all forces between the beads are in balance. One of the biggest mysteries in molecular biology is how this one-dimensional string of amino acid beads can fold itself into the right structure in just a few seconds, when the amino acid string must search over a vast space of possible configurations that the string might conceiv-ably fold-up into. Phrased in computational terms, how can this large collection of strung-together amino acids "compute" a stable thermody-namic structure so quickly?

In 1987, Cyrus Levinthal suggested a solution to this puzzle. His proposal was that nature, through the processes of evolution, has selected for protein sequences having the ability both to fold rapidly and to arrive at a thermodynamically stable structure. Although Levinthal was unable

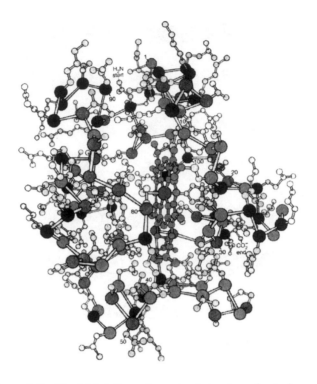

**Figure 2.9**  The folded, three-dimensional structure of *cytochrome c.*

to offer a mechanism by which natural selection could act to single out these amino acid sequences, work reported by Martin Karplus and his co-workers at Harvard in 1994 provides a possible answer to the riddle.

Roughly speaking, Karplus's approach begins by assuming that the process by which the protein folds is governed by an energy function expressing the total energy present in the interactions that can occur between the beads on the protein chain. The chain then folds up in a way that makes the value of this function as small as possible. Karplus then enumerates all possible configurations that the chain can fold into, computing the total energy of each folded form. By examining a large number of random sequences, he found that in the fast-folding sequences there is a wide gap of energy between the folded structure of least energy and the next least energetic one. Slow folding chains, on the other hand, have the least-energy structure, because they are close in energy to nearby structures. We have no room here to go into the details of these experiments, so the interested reader will have to consult the source articles cited in the references. The main point is that to calculate these energy differences between protein chains involves vast amounts of computing resources, again involving the exhaustive enumeration of a large number of individual cases. Again we see the use of the computer as a calculator in the service of science, here in finding the smallest values of a function important in molecular biology. Now let us turn to the geosciences for our last, and by far most familiar, example of the use of the computer as a calculator.

## Weather Forecasting

In a 1904 paper titled "Weather Forecasting as a Problem in Mechanics and Physics," Norwegian meteorologist Vilhelm Bjerknes set the modern tone for the study of the weather. In this first complete formulation of the weather-forecasting problem, Bjerknes anticipated both the experimental and the theoretical components that underlie all numerical weather-forecasting models: (a) knowledge of the initial state of the atmosphere at every point in a spatial grid, and (b) use of the mathematical relations governing the atmospheric dynamics to see how the state at each grid point develops from the states at other points.

The first step in building a numerical weather-forecasting model is to carve up the geographical region of interest by imposing a grid upon it. Figure 2.10 displays the nature of such a grid, along with some of the variables whose values we would like to know in each cell of the grid at each moment of time. These quantities include things like temperature, wind velocity, and barometric pressure. The basic idea is to take the first step of the Bjerknes program and specify the values of each of these variables in each cell at some initial instant. This can be done only by getting actual observations of the weather via satellites, airplane pilots' reports, weather ships, and all the other sources forming the world weather-reporting network.

Having this data, we proceed to Bjerknes's second step and use the mathematical relations to tell us how these values change during the course of time. This computational step involves solving the complicated mathematical equations of the atmosphere. What separates a good prediction from a bad one is the skill with which these two steps—observation and computation—are carried out.

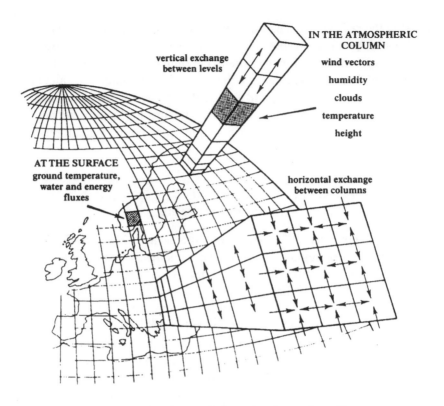

**Figure 2.10** A numerical weather-forecasting grid.

For weather forecasting, the type of model employed is usually called a *general circulation model* (GCM). This is a model that uses a three-dimensional grid like that shown in Figure 2.10, together with a quite detailed set of mathematical relations. In a GCM we have mathematical relations of the following sort:

- *Navier-Stokes equations for horizontal air movement:* These equations describe the movement of air masses in the east-west and north-south directions. This motion is related to the force exerted on the air by the Earth's rotation (the Coriolis force), by the horizontal pressure gradients, by dissipative forces such as friction and turbulence, and by the sources and sinks of momentum.

- *The Navier-Stokes equation for vertical air movement:* This motion is influenced by vertical pressure gradients, gravity, the Earth's rotation, and frictional and turbulence effects.

- *The continuity equation for mass:* This equation expresses the fact that mass is neither created nor destroyed, as well as relating changes in vertical motion to the divergence of the horizontal wind field. Solving this equation gives the vertical motion and predicts the atmospheric pressure at the Earth's surface.
- *The thermodynamic equation:* This equation relates changes in the potential air temperature to the heat supplied by radiation, condensation, and other heat sources.
- *The continuity equation for water:* This a relation for the total water content in all its phases. The equation relates changes in humidity of the air to sources and sinks of moisture.

So much for equations. However, GCMs also include parameters. Here are the main ones:

- *Geophysical:* These are parameters involving the size, rotation, geography, and topography of the Earth.
- *Radiative:* Such parameters refer to the incoming solar radiation and its daily and seasonal variations.
- *Reflectivity:* This category includes quantities related to radiative heat conductive properties of the land surface according to the nature of the soil, vegetation, and snow or ice cover.
- *Oceans:* These are the parameters representing the surface temperatures of the various oceans.

Finally, we come to the punch line: the variables that the GCM actually forecasts. Usually, these are some combination of

- *Winds:* The east-west, north-south, and vertical components of the wind velocities
- *Temperature:* The surface temperature as well as the potential atmospheric temperature
- *Pressure:* The changes in height of the lines of constant pressure (*isobars*) in the atmosphere, along with the surface pressure
- *Water vapor:* The specific humidity and the precipitation.

As a specific example of such a GCM, consider the British Meteorological Office model. It is divided into 15 atmospheric layers starting at the surface of the Earth and rising to a height of 25 kilometers (80,000 feet). In this model, each level (layer) is divided into a horizontal network of points about 150 kilometers apart, giving a total of around 350,000 grid points. Each of these points is assigned values of temperature, pressure, wind, and humidity every 12 hours from observations taken over the entire globe. The equations expressing the change in these quantities are updated numerically, using a computer time step corresponding to 15 minutes of real time. These forecasts are made up to six days into the future. To get some idea of the computing resources required, a one-day forecast involves 100 billion computations and $4\frac{1}{2}$ minutes on a Cyber 205 quasi-supercomputer.

The reason for our spending time on these examples in mathematics, biology and geophysics is mostly to hammer home the point that until very recently the dominant role for the computer has been to act as a number cruncher. The original impetus for building computing machines was to carry out numerical calculations for scientists and engineers, and (with honorable exceptions in fields like artificial intelligence) this historical accident by which we view computers as calculators has dogged the machines ever since. Mercifully those days are finally drawing to a close—and with a vengeance. Today, even school children are taught that the machine is a tool for creating new worlds, not simply something for probing and/or modifying the existing one. To use the computer in this fashion calls for a major intellectual retreading, one centering on a view of the computer as a *symbol processor* rather than as a number cruncher.

Nowadays, we should see the bits and bytes in the machine's memory as representing pictures and alphanumeric characters as with the ASCII code already discussed, rather than seeing these machine symbols simply as numbers. This is the way the computer generates patterns and ideas instead of just passively manipulating numbers. The remainder of our story in this book will be how this new view of the computing machine allows us to see it as a means for *creating* worlds, both electronic miniatures of the world we live in, as well as silicon worlds that we can travel to only with the help of these machines. Because these would-be worlds are the product of the rules embodied in computer programs, let's start the tour by re-examining rules and representations as vehicles of creation.

## The Creative Computer

The modern era of globalist thought in linguistics research was dramatically ushered in with the publication of Noam Chomsky's *Syntactic Structures* in 1957. This electrifying event shifted the focus of linguistics virtually overnight from the collection of specific facts about particular languages—a kind of "verbal botany"—to the identification of this core *universal grammar* from which all human languages get their start. According to the globalists, the universal grammar is something that is biologically present in the mind of all normal children as part of their genetic birthright. In addition to proposing the idea of a universal grammar forming the abstract structure upon which all languages are built, Chomsky also put forth the then radical notion that the grammar of each particular language must be *generative.* By this he meant that the grammar of the language must be a set of rules capable of generating all the well-formed (that is, grammatical) sentences of the language and none of the ill-formed ones. Thus, Chomsky's claim is that even such a creative phenomenon as natural language is the end result of following a set of rules.

The universal grammar characterizes the abstract syntax of language, independent of the peculiarities and idiosyncrasies present in a given human speech community. Thus the basic outline common to every language is coded into a child's genes, with his or her linguistic environment then filling in the details like vocabulary, accent, and word order pertinent to the language being learned. Although the universal grammar allows the learning of any human language, it imposes rather narrow limits on the possible ways that the rules governing each of its subsystems can interact. For instance, languages like Italian have what is called the *null subject option,* allowing statements such as "(blank) went" instead of "he went" or "she went." English has passed up this option. It is the collection of such options that constitutes the boundaries of the universal grammar. However, the grammatical options cannot be chosen freely; the options are interconnected, so that a choice at one level constrains what can be done further down the line. It is also of critical importance to observe that the universal grammar says nothing about the lexical facts of a language, but only about the form of the lexicon. Thus word categories such as nouns and verbs are absent from the universal grammar. But it does contain principles about the assignment of things like semantic roles, cases, voices, and so forth.

What Chomsky's theory says is that sentence structures in different languages show parallels that stem from physiological structures in the human brain. Chomsky bases his generative grammar on a number of rules that he claims are hard-wired into a language "organ" in the brain of every normal human child. This organ contains two types of rules— generative and transformational—which allow us to decompose every sentence. The generative transformations tell us how a sentence decomposes into noun and verb phrases. The verb phrase can then be further decomposed, for example, into a verb and a second noun phrase. In this fashion, the *surface structure* of a sentence is generated and analyzed.

To deal with the so-called *deep structure* (the part that's independent of the specific language under investigation), we must use the transformational rules. With these rules it is possible to separate a sentence from its surface structure, dividing it into a number of simpler sentences. Examples of such transformations are those introduced by relative pronouns, subordinate clause transformations, and transformations from the active to the passive voice. Figure 2.11 shows how the combination of generative and transformational rules can transform the sentences, "These explanations bore the reader. The reader will put aside the book. We expect it." into the single sentence, "We expect that the reader who is bored by these explanations will put aside the book." This example serves to illustrate how a handful of simple rules applied to primitive, "atomic" sentences or phrases can generate a rich linguistic edifice.

The key implication of Chomsky's work for us is that it shows the rule-based nature of all human language. If Chomsky is even approximately correct in his claim that language is a generative/transformational process, then any thought that can be expressed in any language can be built up by applying the appropriate rules from a small number of basic templates common to all languages. A large part of the Chomskyian linguistic program focuses on actually finding these rules and validating their use. Although the jury is still out on whether language is indeed solely a rule-based phenomena peculiar to the human brain, enough has been learned in support of this claim that it can be taken seriously as a starting point for understanding the human language capacity. For our concerns here the key element of Chomsky's work is that it offers evidence for the bold claim that all phenomena—language or otherwise— can be seen as the working out of a set of rules. To add ammunition to this adventurous assertion upon which all the would-be worlds

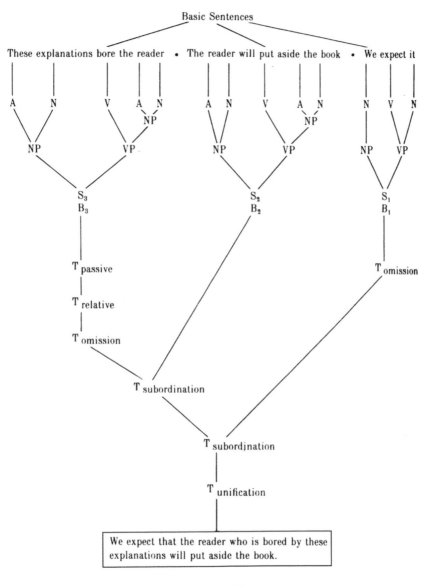

**Figure 2.11**   Chomsky's generative and transformational rules for three sentences.

discussed in this book are based, let's look at two areas closely related to language in which human creativity has been most apparent, music and art, showing how they too can be seen as rule-based processes.

## Tonality and the Structure of Western Music

For the past 300 years Western music has been based on 12 tones or notes, which can be thought of as the 12 hours on the face of a clock. With this image in mind, the pitch interval between two notes is measured in semitones, so that the gap between tones 1 and 2 on the clock face is one semitone. In this way, the distance between any two notes can be measured by the number of semitones separating them.

By moving forward 12 semitones, we go around the clock exactly once, arriving at the same note from which we began. Such a loop is called an *octave*. But just as 1 o'clock on Monday is not the same as 1 o'clock on Tuesday, two notes an octave apart are also different. They vibrate at frequencies that differ by a factor of 2, a relationship that was known at least as far back as the time of Pythagoras in ancient Greece.

Studies of the intervals between notes has turned up the perhaps surprising fact that some intervals just sound better to the human ear than others. In particular, an especially harmonious interval upon which almost all modern Western music is based is one consisting of seven semitones. So, for instance, if we start with the tone 1 and count seven semitones (or hours on the clock face), we arrive at the note 8. Counting seven semitones from there, we come to the note 3. Continuing in this fashion, we generate the sequence of notes 1, 8, 3, 10, 5, 12, 7, 2, 9, 4, 11, and 6. Thus, each of the 12 notes is visited once in this sequence before returning to any note a second time. Now counting forward seven "hours" (semitones) from 6, we arrive back at the starting note 1. This sequence of 12 notes, each separated by seven semitones from its neighbors, is called the *circle of fifths*. Converting this sequence into actual musical notes, if the starting tone 1 is middle C, then the circle of fifths starting at middle C is

C, G, D, A, E, B, F♯, C♯, G♯, D♯, A♯, E♯.

With the development of harmony in the sixteenth century, certain intervals and combinations of notes acquired a preeminent position that they retain to this day. Of special importance are the *major* and *minor scales,* each of which consist of seven notes. These two scales, together with the relationship of fifths, have formed the basis for tonal music for the past 300 years.

The major scale is characterized by the semitone intervals +2, +2, +1, +2, +2, +2. So, if we start with tone 1 and move in these intervals around the clock face, we obtain the major scale beginning with 1 as 1, 3, 5, 6, 8, 10, 12, or in musical notes, C, D, E, F, G, A, and B. Because a scale can begin on any of the notes 1 through 12, there are a total of 12 major scales. Of special importance is what's called a *fifth,* which is a scale that begins seven semitones from the start of a major scale. So, if the scale begins with note 1 as above, the fifth is the scale beginning with the note 8, since 8 is seven "ticks" of the clock from 1. Starting with this note and following the intervals of the major scale, we obtain the major scale 8, 10, 12, 1, 3, 5, 7, or in notes, G, A, B, C, D, E, F♯. Notice that all the notes of the first scale can be found in the second, with the exception of note 6. Similarly, all the notes of the second scale appear in the first except for note 7. This is a general property of scales and the interval of the fifth: There is only a single note difference between the scale of any given note and the scale starting on the fifth of that note.

It's not too great an exaggeration to say that for the past five centuries the history of Western music has been the history of an exploration of the possibilities for expression offered by the special characteristics of the interval of the fifth. The use of the interval of the fifth and its variants became known as *tonality,* and forms the basis for structuring music. It is a way of expressing the relationship of the notes of a particular scale to each other. Whereas keys and modulation shapes the form of piece of music, tonality organizes the overall flow of the music.

What this wildly abbreviated history of musical structure and form tells us is that there are a very small number of rules underlying almost all of traditional Western music. These include the idea of 12 basic semitones, major and minor scales expressed as special intervals between seven of these semitones, and the idea of the fifth. By varying the starting tone, and making re-arrangements, like reversing the sequence of tones and mirroring them, one can lay bare the underlying structure of virtually any piece of music created in the period from 1400–1900. With these ideas as background, let us consider briefly the musical revolution sparked off by the Viennese composer Arnold Schoenberg, who saw an entirely new set of rules for composition.

## Twelve-Tone Music

By 1917, Arnold Schoenberg found himself unable to create new compositions. His music, which was founded on traditional notions of tonality, was being driven primarily by direct emotional expression without being based upon classical musical structures like scales and harmony. Eventually even Schoenberg's emotions could no longer sustain his composition, and he published no music of any kind for a number of years.

In 1923, however, Schoenberg returned to the musical arena with a new idea. Instead of the traditional tonality, he proposed using a method of composition that used all 12 notes of the scale rather than the traditional seven notes of the major and minor scales. In Schoenberg's method, each of the 12 tones of the scale is of equal importance. He further suggested that the 12 notes could be ordered in a series that could be regarded as a theme and used as the basis for a composition. He went on to propose that the 12 notes be ordered so that each note occurs exactly once in the series. Thus, in a piece of music the tones would be used in the order in which they occur in the series, and they would then be played over and over again.

Instead of using tonality as a method to organize the flow of music in a piece, Schoenberg used the same series of notes repeatedly as a means for unifying a work. In this process the role of each note is determined by its position within a series, and the role of the appearance of given series is set by its position within the overall structure of the piece, which consists of several such series. The method of composition pioneered by Schoenberg is called *serialism,* and forms the basis for much of what has been termed *modern* music. Let's take a little closer look at how these new rules of composition work.

Recall that the circle of fifths exemplifies the very essence of tonal music, consisting of the series of 12 tones 1, 8, 3, 10, 5, 12, 7, 2, 9, 4, 11, and 6. A serial composition using the circle of fifths would repeat this sequence over and over. But a serial-based composition of the Schoenberg style represents a new musical language in which the intervals between tones is not so simple as the constant seven semitones of the circle of fifths.

To illustrate serialism à la Schoenberg, consider his *Opus 23,* which uses the series 2, 10, 12, 8, 9, 7, 11, 3, 5, 4, 1, 6. Looking at this series, we find that the intervals are not the constant +7 of the circle of fifths, but rather +8, +2, –4, +1, –2, +4, –8, +2, –1, –3 and +5. Using the clock face image for the 12 tones, this set of different, that is, *nonconstant,* intervals is depicted in Figure 2.12.

At first, it might seem that listening to the same 12 tones repeated over and over again would quickly become rather boring. But Schoenberg's compositional scheme allows for many variations, enabling the composer to add novelty and surprise to the composition. For example, the notes can be played forward, backward, upside-down or even backward and upside-down (inversion), all techniques used earlier in tonal music by composers like Bach and Beethoven. Moreover, the series can be transposed to start on a different note while maintaining the same intervals between the notes. As an illustration of some of these transformations, here is the Schoenberg series for *Opus 23* in upside-down form: 2, 6, 4, 8, 7, 9, 5, 1, 11, 12, 3, 10, 1. This comes from determining the series needed to result in an exchange of the − and + signs in the intervals for *Opus 23*.

This analysis of Schoenberg's atonal scheme, along with our earlier discussion of tonal music, shows that all music can be viewed as a

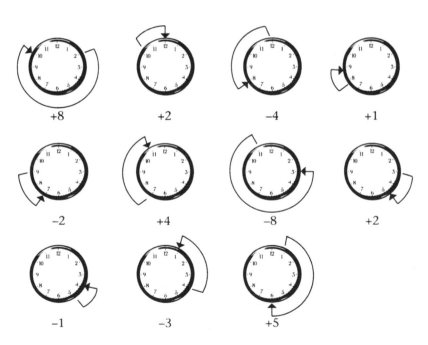

**Figure 2.12**   Intervals in the series of *Opus 23*.

formal abstract system. Even music that seems random has an underlying system. In short, as long as one regards music as a means of communication, there must be a set of rules shared by the composer and the listener that makes the communication possible. There is no reason, in principle, why these rules could not be used by a computing machine to compose music in the style of Bach or Bacharach. In fact, modern composers like Karlheinz Stockhausen, Pierre Boulez, and Gottfried Michael Koenig, to name but a few, have done exactly this as the interested reader can see by consulting the material cited in the bibliography. Now let's see how some of the same rule-based notions have been employed in the visual, rather than aural, arts.

## Evolutionary Art

As we saw earlier in the growth of biomorphs, procedures based on the evolutionary principles of mutation, reproduction, and selection can serve to generate an extraordinary variety of forms. These forms need not be confined to biological structures never before seen on the surface of the earth, either. They may instead lead to completely original artistic works. Let's examine one way to do this.

A few years ago, computer scientist Karl Sims had the idea of regarding expressions in the LISP programming language as genotypes in an evolutionary process. When executed on the computer, the result can then be thought of as the phenotype generated by the LISP expression. Sims's goal was to create a process of artificial evolution using these symbolic expressions. In this process, the LISP expressions open up the opportunity for the emergence of a genuinely new developmental rule or parameter value beyond the boundaries of what may have been set by the programmer at the outset of the experiment. It's not really important for us to know the exact meaning of these LISP expressions, other than that each of them takes a specific number of arguments and returns an image of black-and-white or color values for each pixel in the square. Nevertheless, it is of interest to examine a few of the expressions Sims used just to get a feel for what he had in mind with these experiments.

In Sims's work, the LISP expressions could be formed of combinations of any of the following common LISP functions:

1. *X*
2. *Y*
3. *(abs X)*
4. *(mod X (abs Y))*
5. *(and XY)*
6. *(bw-noise .2 2)*
7. *(color-noise .1 2)*
8. *(grad-direction(bw-noise .15 2).0 .0)*
9. *(warped-color-noise(∗X .2) Y .1 2)*

Figure 2.13 shows the type of image each of these functions produces from an initially black square, where the functions 1–9 are read left to right, top to bottom.

Sims began his experiments by creating a LISP expression that combined a random number of these functions. Such an expression was then translated by a LISP interpreter into a graphic image, the phenotype associated with this symbolic genotype. Since LISP expressions can be written as tree structures, the mutation of such an expression proceeds by traversing the tree, node-by-node, and applying one or another mutation schemes at each node. A typical such scheme might say that if the node is a function like *(abs X),* it might mutate into a different function

**Figure 2.13**  The output from the simple LISP functions 1–9 (see Plate IV).

like, for instance, *(cos Y)*. In addition, symbolic expressions can be reproduced with "sexual combination" by combining the parent expressions in various ways. Figure 2.14 shows the effect of 19 mutations of this sort on a parent in the upper left-hand corner. This figure shows only the surface image of a three-dimensional structure that Sims created by adding a volume texture operation that calculates color values for each point in three-dimensional space.

By starting with randomly generated genomes and applying a variety of types of mutations, Sims played the role of Father Nature, selecting those mutations that would be allowed to live on to the next generation. After anywhere from 5 to 20 generations, a remarkable set of graphic images emerged. Figure 2.15(a–c) shows a small sample of Sims's art gallery, along with the LISP genotypes that coded for these pictures. If you're wondering why there is no genotype displayed for part c of the figure, it is because this phenotype was created before Sims added a genotype-saving subroutine to his program. Part c is what one might call an extinct species in this world of evolutionary art forms. It's a point worth pondering to note here how 186 characters of the alphabet can code for a complicated artistic object like part b of the figure. We'll look into this issue of system complexity in more detail in the next chapter.

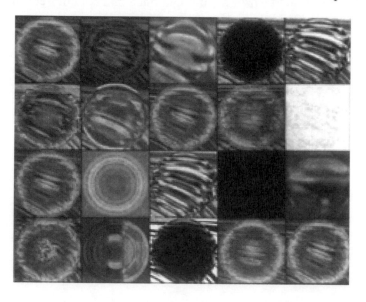

**Figure 2.14**   A parent with 19 mutations (see Plate V).

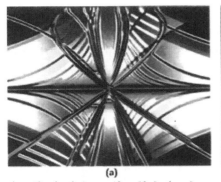

**(a)**

(round(log(+y (color-grad(round(+abs (round
(log(+y(color-grad(round(+y(log(invert y) 15.5))
x)3.1 1.86#(0.95 0.7 0.59) 1.35))0.19)x))(log
(invert y)15.5))x)3.1 1.9#(0.95(0.7 0.35)1.35))
0.19)x)

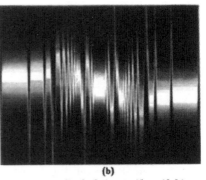

**(b)**

(rotate-vector(log(+y(color-grad(round(+(abs
(round(log #(0.01 0.67 0.86)0.19) x))(hsv-to-
rgb(bump(if x 10.7 y)#(0.94 0.01 0.4)0.78#(
0.18 0.28 0.58)#(0.4 0.92 0.58)10.6 0.23
0.91)))x)3.1 1.93#(0.95 0.7 0.35)3.03))-0.03)
x#(0.76 0.08 0.24))

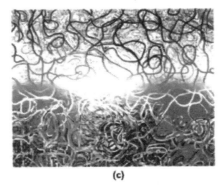

**(c)**

**Figure 2.15** Evolved phenotypes and their corresponding genotypes (see Plate VI).

The foregoing examples of how rules govern communication in language, music, and art suggest that there is no real barrier to having such rules embodied in a computer, thereby opening the possibility of machine creativity. Computers can be creators as well as calculators. We will see much more of this in the pages that follow. For now, let's return to our main theme: the generation of surrogate worlds inside our computing machines by simulation.

# Worlds in Silico

In the mid-1980s, a simulation called *Balance of Power* burst onto the computer gaming scene, giving amateur global strategists and State Department mandarins alike the opportunity to test their skills at keeping the world in one piece. In addition to its unparalleled capacity for creating realistic scenarios and international crises, *Balance of Power* was distinguished by the fact that its designer, Chris Crawford, actually wrote a book about the simulation, which exposed the game's innards for anyone who wished to look inside and see what made it tick. By way of introducing the idea of a simulation, we will take a short look into the workings of the game ourselves.

*Balance of Power* is a game about geopolitics in the modern age. At the time it was put together in the mid 1980s, the Cold War had not yet ended, so the two players in the game are the American president and the general secretary of the Soviet Union, each of whom has the goal of enhancing their country's prestige, without touching off a nuclear exchange. Prestige involves the extent to which your country is liked and respected by other countries; the amount of prestige accrued from being liked by a particular country is proportional to the degree of power that country wields. Your goal is to win lots of powerful friends and have a small number of weak enemies.

As the daily newspaper shows, the stability of almost every nation can be threatened by different geopolitical events. In some cases insurgencies arise to challenge an existing regime by military action. Similarly, coup d'etats, military or otherwise, can sweep a regime out of power literally overnight. In another direction, diplomatic intimidation can induce a nation to "Finlandize" to a superpower; in other words, by taking an accommodating position toward the more powerful nation the intimidated nation will stave off an attack.

The decisions you make to enhance your prestige are all based on one or another of these ways to stabilize potential friends or destabilize existing enemies. For instance, if an unfriendly government is fighting a battle against guerrillas, you might want to give weapons to the insurgents or even send in troops to intervene for the rebels. If the insurgents succeed in ousting the current regime, their leaders will then reward your help with friendly relations, thereby enhancing your prestige.

Of course, your computer opponent can take similar actions to topple your friends. To defend your friends you might want to make weapons shipments to them or, perhaps, try to soothe domestic unrest with economic aid that may help prevent a *coup d'etat* against the friendly regime.

In *Balance of Power*, you are free to take any of these sorts of policy actions anywhere in the world and so is your opponent. However, every move you make is subject to the tacit approval of your opponent, in the sense that if you make a decision your opponent finds objectionable, he (the computer) might demand that you revoke your action. This then provokes a crisis, in which you can either accept the demand and back down, thereby losing prestige, or you may stand firm and reject the demand and thus escalate the crisis. In the latter case, the onus moves back to your opponent who must then choose to either back down or to escalate. This process of escalation or retreat continues until either one side backs down, thus losing prestige, or the crisis escalates to "DefCon 1," the level of international tension at which the missiles start to fly. At this point the world is destroyed in a nuclear Armageddon and both sides lose.

To help the player in deciding what actions to take and the likely reactions of his or her opponent, *Balance of Power* provides numerous maps and databases of economic, social, political, and military facts. An example of such a visual aid is the map shown in Figure 2.16, which depicts the various types of insurgencies taking place in the world at a given moment in time (in this case the year 1986). Taken together, this collection of information provides a snapshot of the state of the world: insurgencies, domestic unrest, diplomatic alignments, and many other factors characteristic of each country of the world. The end result is a simulation that is an admirable compromise between being complex and rich enough to capture the essence of the compromises necessary in geopolitical decision making and being streamlined enough to be playable in a reasonable amount of time on a home computer.

*Balance of Power* is an example of what might be termed a top-down simulation, that is, one in which the basic variables determining the outcome of decisions are aggregated quantities rather than the actions of individual agents. To see what this means, consider a *coup d'etat*. In the real world this is often the result of a combination of the collective discontent of a substantial fraction of the populace, coupled with the individual decision(s) of a small group of powerful military and/or political actors. *Balance of Power*, however, does not follow the detailed actions of individuals (other than the U.S. president and the USSR general secretary). In the simulation the decision about whether a coup is attempted is determined by a simple aggregated relationship like, "If government

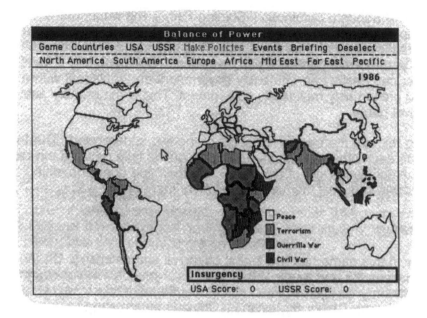

**Figure 2.16**  Insurgencies around the world in 1986.

popularity is less than the USA destabilization plus USSR destabilization, then initiate a coup." Here the variables of government popularity, USA destabilization and USSR destabilization are, in turn, determined by other aggregated variables such as consumer spending per capita, gross national product, population, and economic aid. Similar remarks apply to other key quantities like insurgency and Finlandization.

Use of aggregated variables in a simulation raises the same questions discussed in the opening chapter about fidelity of a model. Just how realistically can a simulation like *Balance of Power* capture the dynamics of geopolitical processes? To get a feel for this question, consider a low-cost flight simulator of the type used for entertainment on personal computers. The fundamental factors calculated by such a program—lift, altitude, air speed, and so forth—are no different from those found in a flight simulator used by airline companies and the military to train their pilots. So how does this multimillion-dollar simulator differ from the one found in your home computer?

The answer lies in Chris Crawford's phrase, "verisimilitude of detail." For example, if your simulated aircraft is moving at an altitude of

10,000 feet, with an angle of attack of 10 degrees and an air speed of 200 knots, the microcomputer simulator does not need to calculate the resultant lift very accurately. A reasonable approximation is all that's required to give the right "feel" for the situation and to make the simulation good enough for entertainment purposes. By way of contrast, the professional simulator had better calculate the lift *very* accurately. If a pilot training on the simulator learns the wrong response to this situation because the lift was only roughly calculated, he or she may repeat this mistaken response in the same situation in real life—possibly with tragic consequences. So a certain level of accuracy in detail is needed in order to take the simulation seriously.

Another difference between a crude model of a situation and simulation comes from the model exaggerating certain features of the real situation. For example, in *Balance of Power* the players expect a clear conflict to emerge. But while conflicts are part and parcel of real life, we have developed many social inhibitions and psychological constraints to soften these conflicts and solve them without resorting to direct confrontation. Thus, any model like *Balance of Power* that strives for commercial success has to accentuate the conflict and remove the inhibitions existing in real life.

Finally, the realism of a simulation is measured by the realism of the process, as opposed to the realism of the data. Most people tend to think that what's important is the realism of the data. For instance, they ask if the Gross National Product is given correctly or if the number of TV sets per capita is right. But data is not the most important factor in assessing realism; rather, it is process. The actual GNP of Australia is less important in a game like *Balance of Power* than the way that GNP changes over the course of time. If our descendants look back 100 years from now, they will see our squabbles with Iraq as so much irrelevant nonsense. But the same principles, the same processes that govern our relationship with Iraq will still be in force. The point is that you cannot interact with a fact; it is a passive element of either the real or the simulated world. But you can interact with a process, and ultimately learn about it. As Chris Crawford put it, "computers are not *data* processors but data *processors.*"

So, by these criteria a top-down simulation like *Balance of Power* has only a moderate level of verisimilitude of detail. It is strong on process but relatively weak on the kind of fine-grained minutiae that may make a difference if you want to use the model to make decisions about real-world events. The difference is akin to the difference between a blueprint of a house and a painting. A painting gives the emotional impact of the house; a blueprint, on the other hand, tells the plumber exactly where to put the sink. But this does not mean that a top-down simulation is a mere low-fidelity version of the "real thing." Instead, it is a simulation that focuses on broader, less quantifiable aspects of the real-world situation. After all, you wouldn't use a painting as the basis for building a house; and, by the same token, you wouldn't use a blueprint to convey an emotional feeling about a house where you had spent your youth. Top-down simulations and their bottom-up counterparts attempt to convey entirely different messages: Bottom-up tries to give detailed technical information; top-down communicates something closer to an artistic impression. Let us turn our attention to a specific bottom-up effort in order to see concretely the difference between the two types of simulation.

## Bubbles, Booms, and Busts

In a little-known dissertation at the University of Paris in 1900, Louis Bachelier laid down the mathematical foundations of the modern academic theory of finance. According to Bachelier, price movements on speculative markets are random, driven solely by new information coming into the market. But because this information itself is unpredictable (otherwise, it would not be information), prices go up or down in a purely random fashion. This argument leads to the so-called efficient-market hypothesis, which in its most common form states that publicly available information is instantly factored into the price of any security; hence, trading schemes based upon past price movements, volumes of transactions, and the like are doomed to fail. In short, there cannot exist a technical trading method based on this kind of information that will consistently outperform a broad market average like the S&P 500. Moreover, in the world of perfectly rational traders envisioned by Bachelier, trading volume should be low, there should be no room for psychology or market moods, and neither crashes nor price bubbles can occur. But a funny thing happened on the way from the lecture hall down to Wall Street.

Traders on the floor of the exchange do not seem to have gotten the message from Paris that technical trading is impossible. Instead, they act as if they believe the market is forecastable, and that methods based on technical indicators and instinct can systematically turn profits far in excess of what could be earned by simply buying and holding a market-basket of stocks like the 30 firms making up the Dow Jones Industrial Average. According to these traders, the market is a complex entity, with anthropomorphic moods and swings. To make money, you simply have to get in tune with the psychology of the market and make your move. So how is one to test these two mutually contradictory views of financial markets? Are the academics like Bachelier right? Is the market nothing more than the flip of a fair coin? Or is it the traders who have the correct view, making their living by putting their money (and yours!) where their mouths are?

Until recently, it wasn't really possible to put these competing theories of market behavior to the test, mostly because academic exper-imentation with, say, trading rules, interest rates and the like on the New York Stock Exchange was frowned up by the Securities and Exchange Commission, not to mention investors and traders around the world. It just was not feasible to perform controllable, repeatable experiments on the market, tests of the type needed to scientifically check out various hypotheses about how prices move in response to various circumstances. Those bad, old days are over.

### A Surrogate Market

In the fall of 1987, W. Brian Arthur, an economist from Stanford, and John Holland, a computer scientist from the University of Michigan, were sharing a house in Santa Fe while both were visiting the Santa Fe Institute. During endless hours of evening conversations, over numerous bottles of beer, Arthur and Holland hit upon the idea of creating an artificial stock market inside a computer, one that could be used to answer a number of questions that people in finance had wondered and worried about for decades. Among those questions were the following:

- Does the average price of a stock settle down to its so-called *fun-damental value*—the value determined by the discounted stream of dividends that one can expect to receive by holding the stock indefinitely?

- Is it possible to concoct technical trading schemes that systematically turn a profit greater than a simple buy-and-hold strategy?

- Does the market eventually settle into a fixed pattern of buying and selling? In other words, does it reach "stationarity"?

- Would a rich "ecology" of trading rules and price movements emerge in the market?

Arthur and Holland knew that the conventional wisdom of finance argued that today's price of a stock was simply the discounted *expectation* of tomorrow's price plus dividend, given the information available about the stock today. This theoretical price-setting procedure is based on the assumption that there is an objective way to use today's information to form this expectation. But this information typically consists of past prices, trading volumes, economic indicators, and the like. So there may be *many* perfectly defensible ways based on many different assumptions to statistically process this information in order to forecast tomorrow's price. For example, we could say that tomorrow's price will equal today's price. Or we might predict that the new price will be today's price divided by the dividend rate. And so on and so forth.

The simple observation that there is no single, best way to process information led Arthur and Holland to the not-very-surprising conclusion that deductive methods for forecasting prices are, at best, an academic fiction. As soon as you admit the possibility that not all traders in the market arrive at their forecasts in the same way, the deductive approach of classical finance theory, which relies upon following a *fixed* set of rules to determine tomorrow's price, begins to break down. So a trader must make assumptions about how other investors form expectations and how they behave. He or she must try to psyche out the market. But this leads to a world of *subjective* beliefs and to beliefs about those beliefs. In short, it leads to a world of induction in which we generalize rules from specific observations rather than one of deduction.

In order to answer the aforementioned questions, Arthur and Holland recruited physicist Richard Palmer, finance theorist Blake LeBaron, and real-world market trader Paul Tayler to help them construct their electronic market, where they could, in effect, play god by manipulating trader's strategies, market parameters, and all the other things that cannot be done with real exchanges. This surrogate market consists of

a. a fixed amount of stock in a single company;

b. a number of traders (computer programs) that can trade shares of this stock at each time period;

c. a specialist (another computer program) who sets the stock price from inside the market by observing market supply and demand and matching orders to buy and to sell, possibly by taking or giving shares from his own supply to facilitate a trade;

d. an outside investment (bonds) in which traders can place money at a varying rate of interest;

e. a dividend stream for the stock that follows a random pattern.

As for the traders, the model assumes that they each summarize recent market activity by a collection of descriptors, which involve verbal characterization like, "the price has gone up every day for the past week," or "the price is higher than the fundamental value," or "the trading volume is high." Let us label these descriptors $A$, $B$, $C$, and so on. In terms of the descriptors, the traders decide whether to buy or sell by rules of the form: "If the market fulfills conditions $A$, $B$, and $C$, then buy, but if conditions $D$, $G$, $S$, and $K$ are fulfilled, then hold." Each trader has a collection of such rules, and acts on only one rule at any given time period. This rule is the one that the trader views as his or her currently most accurate rule.

As buying and selling goes on in the market, the traders can reevaluate their different rules in two different ways: (a) by assigning higher probability of triggering a given rule that has proved profitable in the past, and/or (b) by recombining successful rules to form new ones that can then be tested in the market. This latter process is carried out by use of what is called a genetic algorithm, which mimics the way nature combines the genetic pattern of males and females of a species to form a new genome that is a combination of those from the two parents.

A run of such a simulation involves initially assigning sets of predictors to the traders at random, and then beginning the simulation with a particular history of stock prices, interest rates, and dividends. The traders then randomly choose one of their rules and use it to start the buying-and-selling process. As a result of what happens on the first round of trading, the traders modify their estimate of the goodness of their collection of rules, generate new rules (possibly), and then choose the best rule for the next round of trading. And so the process goes, period after period, buying, selling, placing money in bonds, modifying and generating rules, estimating how good the rules are, and, in general, acting in the same way that traders act in real financial markets.

A typical moment in this artificial market is displayed in Figure 2.17. Moving clockwise from the upper left, the first window shows the time history of the stock price and dividend, where the current price of the stock is the black line and the top of the grey region is the current fundamental value. The region where the black line is much greater than the height of the grey region represents a price bubble, whereas the market has crashed in the region where the black line sinks far below the grey. The upper right window is the current relative wealth of the various traders, and the lower right window displays their current level of stock holdings. The lower left window shows the trading volume, where grey is the number of shares offered for sale and black is the number of shares that traders have offered to buy. The total number of trades possible is then the smaller of these two quantities, because for every share purchased there must be one share available for sale. The various buttons on the screen are for parameters of the market that can be set by the experimenter.

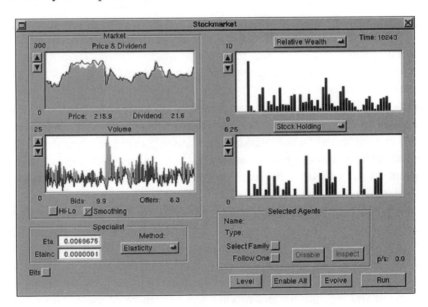

**Figure 2.17** A frozen moment in the surrogate stock market.

After many time periods of trading and modification of the traders' decision rules, what emerges is a kind of ecology of predictors, with different traders employing different rules to make their decisions. Furthermore, it is observed that the stock price always settles down to a random fluctuation about its fundamental value. However, within these fluctuations a very rich behavior is seen: price bubbles and crashes, market moods, overreactions to price movements, and all the other things associated with speculative markets in the real world.

Also as in real markets, the population of predictors in the artificial market continually evolves, showing no evidence of settling down to a single best predictor for all occasions. Rather, the optimal way to proceed at any time depends critically upon what everyone else is doing at that time. In addition, we see mutually reinforcing trend-following or technical-analysis-like rules appearing in the predictor population. These are often new rules that have emerged as the market unfolds. Let's take a quick look at this business of emergent phenomenon, the most important single aspect of these kinds of bottom-up simulations.

## The Emergence of "Emergence"

On Monday, October 19, 1987, the New York Stock Exchange (NYSE) suffered its worst day in history when the value of all stocks on the Exchange declined by an astounding 22.6 percent. Shortly after this calamitous crash, to prevent the kind of panic seen during the Black Monday selling, the Board of Governors of the NYSE instituted so-called circuit-breaker rules prohibiting certain types of trades when the Dow Jones Industrial Average has risen or declined more than 50 points from its previous close. At the time these rules were introduced, it was very unclear whether they would have the desired effect of causing investors to pause and think about their actions rather than blindly leaping in to a turbulent market and following the crowd. There was considerable concern on the part of many market participants that these rules might well exacerbate a disorderly market rather than calm it. As history has shown, these fears were unfounded, because the circuit-breaker rules do indeed seem to have a settling effect on overexcited markets. Could anyone have actually known *in advance* that this would be the case? The short answer is no. Imposition of the circuit-breaker rules was simply a calculated risk on the part of the governors of the Exchange.

Financial markets are perfect examples of complex systems. The feature that distinguishes a complex from a simple system more than any other is a display of strange, surprising and just plain counterintuitive behavior. More often than not, this surprising behavior is attributable to a phenomenon termed *emergence,* which is just system-theoretic jargon for an overall system behavior that comes out of the interaction of many participants—behavior that cannot be predicted or even envisioned from a knowledge of what each component of the system does in isolation. A simple example from chemistry is everyday tap water. Water is formed from two molecules of hydrogen and one of oxygen, both rather flammable, reactive gases. Yet when the two are combined, the result is a compound that is liquid and nonflammable—both properties that have "emerged" as a result of the interaction of the component parts.

It is the ever-present possibility for unexpected (and often unwanted) emergent behavior that made the circuit-breaker rules imposed on the New York Stock Exchange potentially dangerous. There was no way in 1987 to safely experiment with the market mechanisms, in order to see if the rules would work without running the very real risk of inadvertently imposing trading regulations that might make the markets more unstable rather than less. With microsimulations like that of Arthur & Co., those days are now behind us. In today's high-tech world we can actually create surrogate worlds in our computers that allow us to carry out repeatable, controllable, scientific experiments on such complex systems as stock markets.

Although emergent phenomena are undoubtedly the striking feature tending to separate complex systems from the simple, they are far from the only such properties. When we see microworlds like the aforementioned stock market in operation, we see many other features characteristic of complex systems. Before entering into an extended account of other types of surrogate worlds besides the stock market, it's worth taking time to examine in more details some of these characteristics of complexity.

# CHAPTER

# 3

# The Science of Surprise

## Problems and Paradoxes

A popular toy a number of years ago was a gadget called either the Bouncer or the Space Trapeze, which is shown in Figure 3.1. Although it came in many forms, the Bouncer generally consisted of two light, hollow balls containing magnets. Its axis swung to and fro at the top of a pendulum, whose bob was a heavy driving ball. The regular swing of the driving ball was itself driven by an electromagnet, located in the base of the device and switched by a circuit that sensed each approach of a magnet inside the pendulum bob.

What made the Bouncer interesting as a toy was the fact that as well as rotating inertially and being swung by the pendulum bob, the rod containing the two light balls occasionally bounced because its balls were repelled by a magnet located just above the pendulum. As a result of these various influences, the rod containing the light balls rotated in an irregular fashion, clockwise and counterclockwise, on time scales ranging from several seconds to several weeks (when the battery for the electromagnet in the base ran down).

**Figure 3.1**   The Bouncer.

Viewed as a dynamical system, the Bouncer serves as a concrete example of a "chaos machine," because its behavior illustrates perfectly how small changes in the initial configuration of a system (the initial position and velocity of the arm containing the two balls) can give rise to dramatically different behaviors of the system (subsequent motions of the swinging arm). Thus, the Bouncer shows us one way in which complex behavior can emerge out of a simple cause, in this case instability, whereby small disturbances at one place in the system give rise to large changes elsewhere. However, instability is not the only surprise-generating mechanism in the system analyst's toolkit. Let's look at another, this time in the context of a simple economy.

## Paradoxical Behavior in a Developing Economy

Consider a job-training system in a developing country, whose nonagricultural economy has two kinds of workers and two factories: machinists who work in the job shop and electricians who work for the electrical contractor. Both the job shop and the electrical contractor have the capacity to employ a fixed number of workers, and they try to operate at full capacity. For the sake of illustration, assume that workers leave the work force sufficiently often that the number of workers equals the yearly output of the schools. There are three schools: two small schools, specializing in training machinists and electricians, and a large government school, training an equal number of both kinds of workers. The government trains two workers per dollar, whereas the private schools train one worker per unit of demand. Because these private schools can be more selective in the students they accept, they train their students to twice the productive capacity of the government-trained workers. The government subsidizes the factories so that they will take all workers trained in the government school.

Putting all these remarks and definitions together, we have the following relations:

$$\text{number of machinists} = \text{demand for machinists} + \text{output of government school,}$$

$$\text{number of electricians} = \text{demand for electricians} + \text{output of government school,}$$

$$\text{total productive capacity} = \text{demand for machinists} + \text{demand for electricians} + \text{output of government school.}$$

Assume that the country's economic planners want to control the number of machinists and electricians, as well as the total productive capacity of the economy, so as to achieve a high level of employment along with production. The controllers are the two factories and the government. The government controls the productive capacity via the output of its school, while the job shop controls the number of machinists by the demand for machinists. Similarly, the electrical contractor controls the number of electricians with the demand for electricians.

This situation generates the following paradoxical behavior: Suppose the two factories have been operating at full capacity. The government then increases the output of its school by one unit. In turn, the two factories decrease their demands by one unit each to avoid overflowing. The net effect on total productive capacity of the changes in the two demands is then minus two units. Thus, the overall effect on productive capacity of this single unit increase in the output of the government school is a one unit *decrease* in production. This conclusion is independent of the detailed control strategies employed by the planners, depending only upon the structure of control and the objectives as seen by each controller. So the more trained workers in the workplace, the less the productive capacity—a paradox.

The paradox could be avoided if the government could manipulate either of the demands instead of controlling only the output of its school.

However, the basic problem arises because of the effect of other control actions on the apparent relationship between a controlled variable (the number of machinists, the number of electricians, or the productive capacity) and the corresponding decision variable (the demand for machinists, the demand for electricians, or the output of the government school). The moral of the story is that seemingly elementary systems can give rise to very unexpected (and unpleasant) outcomes if the nature of interactions among the components of the system is not thoroughly understood. Also, the counterintuitive aspects of this situation are not due merely to disproportionalities or random effects. Rather, they are attributable mostly to the interactions among the subsystems and the way these subsystems are connected to each other.

Both the Bouncer and the job-training paradox illustrate the point that the hallmark of complex systems is their capacity to display counterintuitive or just plain surprising behavior. In the case of the Bouncer, its behavior is simply unpredictable because of an almost pathological instability in the dynamics governing the motion of the arm containing the two lightweight balls. On the other hand, the economic paradox, in which increased government spending on job training actually reduces the productive capacity of the economy, is attributable solely to the way information, people, and money are allowed to flow through the economy. What both examples have in common is that they defy our expectations about what the system will do next. In other words, the Bouncer and the economy display the feature that most clearly distinguishes complex systems from those that are simple: They display surprising, counterintuitive behavior. If we want to get a handle on the kinds of behaviors seen in the would-be worlds discussed in the next chapter, it's going to be essential to identify the various surprise-generating mechanisms governing the behavior of complex systems. This is the *sine qua non* for the development of a decent theory of complex systems. We begin this quest by discussing five such mechanisms in brief, and then considering each in greater detail.

## The Fingerprints of the Complex

The vast majority of counterintuitive behaviors shown by complex systems are attributable to some combination of the following five sources: paradoxes/self-reference, instability, uncomputability, connectivity, and emergence. With some justification, we can think of these sources of complexity as *surprise-generating mechanisms,* whose quite different natures each lead to their own characteristic type of surprise. Let's take a quick look at each of these mechanisms before turning to a more detailed consideration of how they act to create complex behavior.

**Paradox.** Paradoxes arise from false assumptions about a system leading to inconsistencies between its observed behavior and our *expectations* of that behavior. Sometimes these situations occur in simple logical or linguistic situations, such as the famous Liar Paradox ("This sentence is false."). In other situations, the paradox comes from the peculiarities of the human visual system, as with the impossible staircase shown in Figure 3.2, or simply from the way in which the parts of a system, like the developing economy discussed in the preceding section, are put together.

**Instability.** Everyday intuition has generally been honed on systems whose behavior is stable with regard to small disturbances, for the obvious reason that unstable systems tend not to persist long enough for us to develop good intuitions about them. Nevertheless, the systems of both nature and humans often display pathologically sensitive behavior to small disturbances, as for example, when stock markets crash in response to seemingly minor economic news about interest rates,

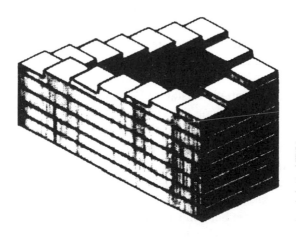

**Figure 3.2**　The impossible staircase.

corporate mergers, or bank failures. Such behaviors occur often enough that they deserve a starring role in our taxonomy of surprise. Here is a simple example illustrating the point.

In Adam Smith's (1723–1790) world of economic processes, a world involving a system of goods and demand for those goods, prices will always tend toward a level at which supply equals demand. Thus, this world postulates some type of negative feedback from the supply/demand relationship to prices, which leads to a level of prices that is stable. This means that any change in prices away from this equilibrium will be resisted by the economy, and that the laws of supply and demand will act to reestablish the equilibrium prices. Recently, maverick economists like W. Brian Arthur of Stanford and the Santa Fe Institute have argued that this is not the way many of the sectors work in the real economy at all. Rather, these economists claim that what we see is *positive* feedback in which the price equilibria are unstable.

For example, when video cassette recorders (VCRs) started becoming a household item some years back, the market began with two competing formats—VHS and Beta—selling at about the same price. By increasing its market share, each of these formats could obtain increasing returns since, for example, large numbers of VHS recorders would encourage video stores to stock more prerecorded tapes in VHS format. This, in turn, would enhance the value of owning a VHS machine, leading more people to buy machines of that format. By this mechanism a small gain in market share could greatly amplify the competitive position of VHS recorders, thus helping that format to further increase its share of the market. This is the characterizing feature of positive feedback—small changes are amplified instead of dying out.

The feature of the VCR market that led to the situation described above is that it was initially unstable. Both VHS and Beta systems were introduced at about the same time and began with approximately equal market shares. The fluctuations of those shares early on were due principally to things like luck and corporate maneuvering. In a positive-feedback environment, these seemingly chance factors eventually tilted the market toward the VHS format until the VHS acquired enough of an advantage to take over virtually the entire market. But it would have been impossible to predict at the outset which of the two systems would ultimately win out. The two systems represented a pair of unstable equilibrium points in competition, so that unpredictable chance factors ended up shifting the balance in favor of VHS. In fact, if the common claim that the Beta format was technically superior holds any water, then the market's choice did not even reflect the best outcome from an economic point of view.

**Uncomputability.** By their very nature, the kinds of behaviors seen in the surrogate worlds presented in the preceding chapters, such as the artificial stock market or *Football Pro '95,* are the end result of following a set of rules. This is because these worlds are embodied in computer programs, which in turn are necessarily just a set of rules telling the machine what bits in its memory array to turn on or off at any given stage of the calculation. By definition, this means that any behavior seen in such worlds is the outcome of following the rules encoded in the program. Although computing machines are *de facto* rule-following devices, there is no *a priori* reason to believe that any of the processes of nature and humans are necessarily rule based. If uncomputable processes do exist in nature—for example, the breaking of waves on a beach or the movement of air masses in the atmosphere—then we could never see these processes manifest themselves in these surrogate worlds. We may well see processes that are close approximations to these uncomputable ones, just as we can approximate an irrational number as closely as we wish by a rational number. However, we will never see the real thing in our computers, if indeed such uncomputable quantities even exist outside the pristine world of mathematics.

To illustrate what is at issue here, the problem of whether the cognitive powers of the human mind can be duplicated by a computing machine revolves about just this question. If our cognitive activity is nothing more than following rules encoded somehow into our neural circuitry, then there is no logical obstacle to constructing a "silicon mind." On the other hand, it has been forcefully argued by some, most recently physicist Roger Penrose, that cognition involves activities that transcend simple rule following. If so, then the workings of the brain can never be captured in a computer program and the technophobes of the world can all rest easier at night. This is because there can then be no set of rules—a computer program—that could faithfully capture *all* the things happening in the cognitive processes of the human brain.

**Connectivity.** What makes a system a system and not simply a collection of elements is the connections and interactions among the individual components of the system, as well as the effect these linkages have on the behavior of the components. For example, it is the inter-relationship between capital and labor that makes an economy. Each component taken separately would just not do. The two must interact for economic activity to take place. As we saw with the example of the developing world economy, complexity and surprise often resides in these connections. Here is another illustration of this point.

Certainly the most famous question of classical celestial mechanics is the *N-Body Problem,* which comes in many forms. One version involves $N$ point masses moving in accordance with Newton's laws of gravitational attraction, and asks if from some set of initial positions and velocities of the particles, there is a finite time in the future at which either two (or more) bodies collide or one (or more) bodies acquires an arbitrarily high energy, and thus flies off to infinity. In the special case when $N = 10$, this is a mathematical formulation of the question, "Is our solar system stable?"

The behavior of two planetary bodies orbiting each other can be written down completely in terms of the elementary functions of mathematics, like powers, roots, sines, cosines, and exponentials. Nevertheless, it turns out to be impossible to combine the solutions of three two-body problems to determine whether a three-body system is stable. Thus, the essence of the Three-Body Problem resides somehow in the way in which *all three* bodies interact. Any approach to the problem that severs even one of the linkages among the bodies destroys the very nature of the problem. Here is a case in which complicated behavior arises as a result of the interactions between relatively simple subsystems.

Incidentally, in a 1988 doctoral dissertation based upon earlier work by Don Saari, Jeff Xia of Northwestern University gave a definitive answer to the general question of the stability of such systems by constructing a five-body system for which one of the bodies does indeed acquire an arbitrarily large velocity after a finite amount of time. This result serves as a counterexample to the idea that perhaps *all n*-body systems are actually stable. The general idea underlying Xia's construction is shown in Figure 3.3, where we see two binary systems and a fifth body that shuttles back and forth between them. By arranging things just right, the single body can be made to move faster and faster between the binaries until at some finite time its velocity exceeds any predefined level.

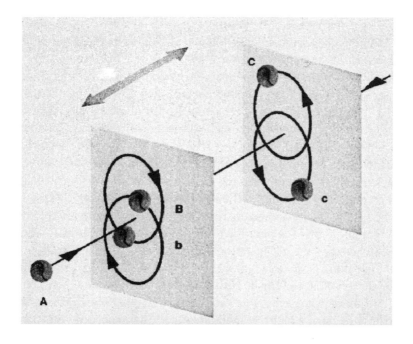

**Figure 3.3**  Xia's solution of the *N-Body Problem.*

This result says nothing about the specific case of our solar system, which is certainly not configured like Xia's example. In fact, the computer simulations of the formation of planetary systems given in chapter 1 suggest that most real planetary systems are very different than the special configuration used by Xia—a good example of the difference between the way a physicist and a mathematician look at the same problem! But Xia's example does suggest that *perhaps* the solar system is not stable, and more importantly offers new tools with which to investigate the matter further.

**Emergence.**  A surprise-generating mechanism dependent on connectivity for its very existence is the phenomenon of emergence. This refers to the way the interactions among system components generates unexpected global system properties not present in any of the subsystems taken individually. We touched on this issue briefly in chapter 2, mentioning how the distinguishing characteristics of water—being a liquid and being nonflammable—are totally different than the properties of its component gases, hydrogen and oxygen.

The difference between complexity arising from emergence and that coming only from connection patterns lies in the nature of the interactions among the various component pieces of the system. For emergence, attention is not simply on whether there is some kind of interaction between the components, but also on the specific nature of that interaction. For instance, connectivity alone would not enable one to distinguish between ordinary tap water involving an interaction between hydrogen and oxygen molecules and heavy water (deuterium), which involves interaction between the same components albeit with an extra neutron thrown in to the mix. Emergence would make this distinction. In practice it is often difficult (and unnecessary) to differentiate between connectivity and emergence, and they are frequently treated as synonymous surprise-generating procedures. A good example of emergence in action is the organizational structure of an ant colony.

Like human societies, ant colonies achieve things that no individual ant could accomplish: Nests are erected and maintained, chambers and tunnels are excavated, and territories are defended. All these activities are carried on by individual ants acting in accord with simple, local information; there is no master ant overseeing the entire colony and broadcasting instructions to the individual workers. Somehow each individual ant processes the partial information available to it in order to decide which of the many possible functional roles it should play in the colony.

Recent work by Stanford biologist Deborah Gordon on harvester ants in southeastern Arizona has shed considerable light on the process by which an ant colony assesses its current needs and assigns a certain number of members to perform a given task. Gordon identifies four distinct tasks an adult harvester-ant worker can perform outside the nest: foraging, patrolling, nest maintenance, and midden work (building and sorting the colony's refuse pile). So it is these different tasks that define the components of the system we call an ant colony, and it is the interaction among ants performing these tasks that gives rise to emergent phenomena in the colony.

One of the most notable interactions is between forager ants and maintenance workers. When Gordon increased nest-maintenance work by piling some toothpicks near the opening of the nest, the number of foragers decreased. Apparently, under these environmental conditions the ants engaged in task switching, with the local decision made by

**Table 3.1**   The main surprise-generating mechanisms.

| Mechanism | Surprise Effect |
|-----------|-----------------|
| Paradoxes | Inconsistent phenomena |
| Instability | Large effects from small changes |
| Uncomputability | Behavior transcends rules |
| Connectivity | Behavior cannot be decomposed into parts |
| Emergence | Self-organizing patterns |

each individual ant determining much of the coordinated behavior of the entire colony. Task allocation depends on two kinds of decisions made by individual ants. First, there is the decision about which task to perform, followed by the decision of whether to be active in this task. As already noted, these decisions are based solely on local information; there is no central decision maker keeping track of the big picture.

Figure 3.4 gives a summary of the task-switching roles in the harvester ant colony. Gordon's work has shown that once an ant becomes a forager, it never switches back to other tasks outside the nest. When a large cleaning chore arises on the surface of the nest, new nest-maintenance workers are recruited from ants working inside the nest, not from workers performing tasks on the outside. When there is a disturbance like an intrusion by foreign ants, nest-maintenance workers will switch tasks to become patrollers. Finally, once an ant is allocated a task outside the nest, it never returns to chores on the inside.

The ant colony example shows how interactions among the various types of ants can give rise to patterns of global work allocation in the colony, patterns that could not be predicted or that could not even arise in any single ant. These patterns are emergent phenomena due solely to the types of interactions among the different tasks.

Table 3.1 gives a summary of the surprise-generating mechanisms we have examined in this section. Now let's take a longer look at how each of them actually operate in practice as well as in principle.

## Illusions of the Mind

In 1947, two Harvard undergraduates, William Burkhart and Theodore Kalin, built the world's first electronic computer, a machine designed

93

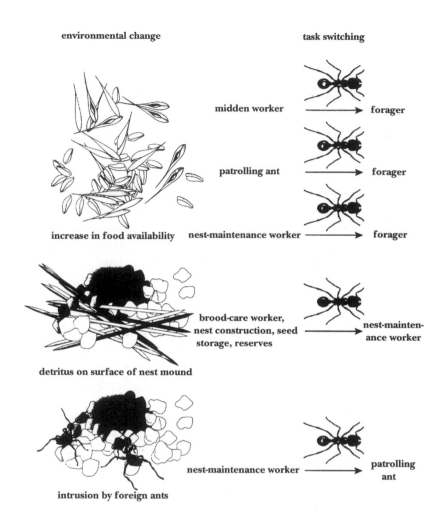

environmental change

task switching

midden worker ⟶ forager

patrolling ant ⟶ forager

increase in food availability    nest-maintenance worker ⟶ forager

brood-care worker, nest construction, seed storage, reserves ⟶ nest-mainten-ance worker

detritus on surface of nest mound

nest-maintenance worker ⟶ patrolling ant

intrusion by foreign ants

**Figure 3.4** Task switching in a harvester ant colony.

solely to solve problems in logic. When they asked their machine to evaluate the Liar Paradox ("This sentence is false.") cited earlier, Kalin reports that the machine went into wild oscillations and made "a hell of a racket," thus providing a mechanical response to the inconsistency inherent in such self-referential statements. The Liar Paradox itself is mostly a philosophical cum linguistic puzzle, whose paradoxical nature is evident almost by inspection. But self-reference, in which statements refer to themselves as in the Liar Paradox, arises in far more subtle ways, often leading to counterintuitive behavior in natural and human systems. Let's look at a legal example to get the flavor of what kind of surprising things can happen when self-reference enters the picture.

The Greek philosopher Protagoras is reported to have taught law to a poor student named Euarthlus on the condition that Euarthlus would pay Protagoras's fee as soon as he won his first case. Upon completing his studies with Protagoras, Euarthlus abandoned the law for a life in politics. After waiting some time to receive his fee, Protagoras finally sued Euarthlus for the fee.

In the court battle, Protagoras argued that if Euarthlus lost the case, then he would have to obey the court and pay the fee. If Euarthlus won the case, then he would have satisfied the terms of the original agreement by winning his first case, and hence would also owe the fee to Protagoras. Thus, Protagoras claimed that whatever the outcome of the lawsuit, Euarthlus would be obliged to pay the fee.

Euarthlus's argument, however, is equally persuasive. He maintained that if he won the case, then the court would have ruled in his favor and he would not have to pay Protagoras. Moreover, if he lost the case, then he would not have to pay Protagoras either, since he would then still not have won his first case.

So who's argument is correct? Basically, neither, since Euarthlus and Protagoras are actually locked into a self-referential paradox of exactly the same sort as the Liar Paradox discussed above. The only difference is that unlike the case of the Liar Paradox, the paradox here does not arise from the chain of arguments given by either litigant taken separately. Rather, it comes from the *combination* of the arguments presented by Euarthlus and Protagoras taken as a whole.

To see the essential ingredients of the Liar Paradox hiding beneath this pair of arguments, it's useful to look at a variant of this type of self-reference invented by the French mathematician Philip Jourdain in 1913. Jourdain distributed calling cards printed on both sides with the messages shown in Figure 3.5. Jourdain's calling card paradox is a variant of a paradox noted by fourteenth-century Venetian philosopher Jean Buridan. According to Buridan, Socrates makes the single statement, "What Plato says is false." Similarly, Plato makes the statement, "What Socrates says must be true." The combination of these two statements immediately leads to the conclusion that what Socrates says must be both true and false at the same time.

## FRONT

> The statement on the other
> side of this card is true.

## BACK

> The statement on the other
> side of this card is false.

**Figure 3.5**  Philip Jourdain's calling card.

What is interesting about these Liar Paradoxes is that the individual statements taken separately make perfectly good sense, but like mixing nitric acid and glycerin, things blow up in your face when the two statements are combined. Self-reference, though, is far from the only way that systems can display paradoxical behavior. Here is another from the theory of social choice.

### Arrow's Impossibility Theorem

Suppose we have a group like a legislative body or a society trying to make a decision on democratic grounds, so that the desires of each individual in the group are weighted equally. Usually each member of the group has his or her preferences or opinions as to what the best thing to do is, so we would like to establish fair and uniform procedures that the group can use to combine individual preferences to reach some sort of consensus. We assume that each member of the group can rank order the various alternatives available. The collection of all such rankings is called the group profile. To avoid degeneracies in the argument, we assume that there are at least three alternatives that each individual must rank in order of how he or she feels about them. For instance, suppose the choices are among flavors of ice cream, say, vanilla, chocolate, and strawberry. Then each person must produce a list saying which flavor is their favorite, least favorite, and intermediate favorite. In my case, that rank order would be $V > S > C$, indicating that I most prefer vanilla, followed by strawberry, with chocolate bringing up the rear.

The problem is to produce a rule for determining a ranking for the group from these individual profiles. Note here that a *ranking* is an ordering of the alternatives that is both *transitive* (if *A* is preferred to *B* and *B* is preferred to *C*, then *A* is preferred to *C*), and *asymmetric* (if *A* is preferred to *B*, then *B* is not preferred to *A*). The question is whether any rule exists that will start with the rankings given by the individuals and produce a single ranking for the group that meets certain minimal requirements for democratic decision making. In the early 1950s, economist Kenneth Arrow showed that there could be no such rule. Let's look at the basis for this somewhat surprising and disturbing conclusion.

Arrow's Impossibility Theorem rests upon the following five axioms, which are intended to reflect conditions that a reasonable aggregation rule should satisfy in any society that fancies itself as being democratic.

**AXIOM I.** *(Universal Admissibility of Individual Preferences.) All possible orderings of alternatives by individuals are allowed. This means that there are no institutions or other agencies that can restrict individuals in their selection of personal preferences.* ■

**AXIOM II.** *(Independence from Irrelevant Alternatives.) If* S *is some subset of the alternatives, and the preferences of individuals change with respect to alternatives* not *in* S, *then the society's ordering for alternatives in* S *does not change.* ■

**AXIOM III.** *(Positive Association of Individual and Social Values.) Suppose society prefers* A *to* B *and people change their minds about other alternatives—but not about* A *and* B. *Then* A *should still be preferred to* B. *In other words, society's decision about whether* A *is better than* B *should not depend on its decision about the relative merits of some other pair of alternatives* X *and* Y. ■

**AXIOM IV.** *(Citizens's Sovereignty.) This means that for any two alternatives* A *and* B, *there must be some group ranking that would* *allow society as a whole to also prefer* A *to* B. *If not, there would then be a pair of alternatives* A *and* B *such that even if every individual preferred* A *to* B, *the group would never prefer* A *to* B. *In this event, the society's ordering between* A *and* B *is* imposed *on the group, and the individuals' preferences do not carry any weight.* ∎

---

**AXIOM V.** *(Nondictatorship.) The society contains no dictator, whose preferences are forced upon the society as a whole.* ∎

Arrow showed that under these assumptions, which almost all observers feel are perfectly reasonable requirements for any democratic method of decision making based on expressing individual preferences by means of voting, there cannot exist a perfect democratic voting system, that is, one in which the majority's choice is always preferred.

The root cause of Arrow's paradox resides in the difference between transitive and nontransitive relations. For example, the relation "is older than" is transitive, since if $A$ is older than $B$ and $B$ is older than $C$, then $A$ is older than $C$. To see that not all relations are transitive, consider the relation "loves." It's clear that if $A$ loves $B$ and $B$ loves $C$, it does not necessarily follow that $A$ loves $C$. In fact, $A$ may *hate* $C$. Arrow proved that any voting system that involves aggregating individual preferences that are transitive, and that satisfies the five axioms given above, is subject to the paradox. Of course, all democratic voting schemes should be transitive, since if each person prefers $A$ to $B$, $B$ to $C$, and $A$ to $C$, then the group as a whole should prefer $A$ to $C$. With the foregoing axioms, however, there is no procedure, that is, a rule, for putting such individual preferences together that will preserve this ordering. Only by dropping one or another of these conditions can the paradox be avoided. Oddly enough, the most commonly proposed resolution of the paradox is to drop Axiom V, the "No Dictator" assumption, and appoint a judge or arbitrator to make a decision when the voting system is unable to produce a clear-cut majority winner.

Finally, look at another type of paradox in which the surprising behavior arises out of the conflict between making a decision on the basis of alternatives, one of which dominates the others, and choosing on the basis of the expected payoffs, in a statistical sense, to be obtained from each alternative.

## The Nomination Paradox

Consider the following situation from the political arena. We have three persons, Albert, Basil, and Clyde, or $A$, $B$, and $C$ for short, who are contemplating entering the race for their political party's nomination for the presidency. Candidate $A$ announces first, saying that he will enter all the primaries. Candidate $B$ then announces that he will defer his decision on whether to enter the primaries, presumably to assess his chances should candidate $C$ also decide to enter the race. The reasoning underlying this decision has been given by game theorist Steven Brams as follows:

> Candidate $C$ will not make a decision on whether to enter the primaries until the last minute—too late for me to organize an effective campaign should I decide to enter. Therefore, in order to determine my own best course of action, I must predict what he will decide to do.
>
> 1. If I predict correctly that $C$ will enter, I should not enter myself but instead let $A$ and $C$ fight it out in the primaries. Candidate $C$ will almost certainly win in most of the primaries, but I would have a 50–50 chance of beating him at the convention.
> 2. If I incorrectly predict that $C$ will enter, I should not enter myself and $A$ will then run unopposed in the primaries. $A$'s victories will then not be very impressive. But because $C$ has better credentials as a compromise candidate than I do, he would have a better than 50–50 chance of beating both $A$ and me at the convention, where his middle-of-the-road position will be most attractive.
> 3. If I incorrectly predict that $C$ will not enter the primaries, and I therefore decide to enter, $C$ as the compromise candidate will almost certainly lose in a three-way primary race, where his middle-of-the-road position will not appeal to the voters.
> 4. If I correctly predict that $C$ will not enter the primaries, and I therefore decide to enter, I can almost certainly beat $A$ in the primaries. With this strong primary support, $C$, even as a compromise candidate, would have less than a 50–50 chance of beating me at the convention.

The payoffs to candidates $B$ and $C$ for these four outcomes are shown in the following payoff matrix:

|  | Candidate B | |
| --- | --- | --- |
|  | **Predict C enters** | **Predict C does not enter** |
| Candidate C — **Enters** | 50–50 chance | Loses |
| Candidate C — **Does not enter** | Chances > 50–50 | Chances < 50–50 |

A first look at these payoffs immediately suggests that there is no problem, because the payoffs associated with candidate $C$ not entering the primaries dominate those received when he enters. So it would seem that the dominance principle argues in favor of candidate $C$ not entering the primaries *regardless* of the prediction made by candidate $B$. But not so fast!

Suppose candidate $C$ had complete confidence in $B$'s ability to predict whether he would enter or not. Then, since he cannot possibly outsmart $B$—because $B$ would know about it and act accordingly in making his prediction—$C$ should enter the primaries. Correctly predicting this, candidate $B$ would not enter, but would wait for the convention to contest the nomination. This pair of choices would lead to a greater than 50–50 chance of winning the nomination for candidate $C$, superior to the less than 50–50 chance he would have if he did not enter and $B$ correctly predicted this choice. So we see that in this extreme case, when candidate $C$ feels that $B$ can infallibly predict what he will do, the dominance principle should be ignored and he should enter the primaries.

This conundrum, which generally goes under the rubric *Newcomb's Paradox,* rests upon whether one should make decisions on the basis of dominance or expected return. We have sketched the basics of the idea here in the context of political decision making to show that the dilemma is not an academic curiosity, but rather can easily arise in real-life situations. For instance, Brams has given a convincing argument that the scenario outlined here for the mythical candidates $A$, $B$, and $C$ fits very closely the situation surrounding the 1968 race for the Democratic presidential nomination, taking Eugene McCarthy as candidate $A$, Robert Kennedy as candidate $B$, and Hubert Humphrey as candidate $C$. For example, outcome 2 supposes that if McCarthy had run unopposed in the primaries, Humphrey could probably have captured the nomination at the Chicago convention. This is a plausible scenario, since even with considerable support from the Kennedy delegates at the convention, McCarthy did in fact lose the nomination to Humphrey. For a discussion of how the other outcomes match up to the 1968 Democratic presidential race, I refer the interested reader to the complete story given in the material by Steven Brams cited in the references.

Now let's turn our attention from surprises coming from paradoxes of different sorts to the kind of counterintuitive behavior arising from instabilities. Instead of logical time bombs, we will direct our attention to how unexpected big things, like the outbreak of warfare or the crash of a financial market, can come about from seemingly small changes, such as a malevolent dictator's indigestion or a rumored increase in unemployment figures.

## A Little Can Be a Lot

Arnold Toynbee's *A Study of History* emphasizes the rise and fall of great civilizations as being a process of challenge and response. In Toynbee's view, a threat to a society from the outside can either strengthen or reduce a civilization's integrity. As a result, that civilization can either dominate nearby cultures or be subjugated by them. To illustrate, let's consider patterns of growth and decline of the Roman Empire, in which *challenge* and *political and economic integrity* are taken as independent causal variables affecting the *cohesion* of the Empire.

Of special interest for us is the path by which the Roman Empire collapsed. Basically, this path involved an ever-increasing level of political and economic unity, coupled with external challenges bringing the Empire to the critical point at which the cohesion of the Empire actually begins to decrease while the challenges continued to grow. Prior to this point, the Empire was able to rise successfully to the various external challenges from invaders and local warlords and maintain a high level of central domination. But at the critical point, the system was poised on the knife-edge of collapse, and it took only a very slight outside challenge for it to be overthrown, thereby rapidly sending the Empire into a state of subjugation.

Such a critical point is an example of an *unstable* point in the dynamics of the rise and fall of the Roman Empire, because it is a point at which small, seemingly insignificant changes in the integrity and external challenges can lead to a discontinuous shift in the observed behavior of the system. On the other hand, points where the Empire is being formed are *stable,* since a small change in either the external challenge or the integrity levels that move the system to nearby points results in only a correspondingly small change in the Empire's dominion.

Further examination of the dynamics of the Roman Empire shows that the reason for its rapid collapse is the rather high level of political and economic integrity that developed in the face of very little, if any, external challenges. This is what set the stage for the sharp decline. However, the very same end result could have occurred gradually if, for example, the Romans had developed their political and economic structures more slowly, along with responding to a parallel set of external threats. In this case, the system could have followed a path resulting in a gradual movement from a state of central domination to one of external subjugation.

Technically speaking, the kind of (in)stability displayed in the collapse of the Roman Empire is called *structural stability,* since the changes in either of the variables, *challenge* or *integrity,* is actually a change in the dynamic process underlying movement from domination to/from subjugation. There is another type of instability, one associated more with a single system operating in its far-from-equilibrium regime than a family of systems and their respective equilibria. This is the kind of stability often associated with so-called chaotic systems.

### The Saltwater Taffy Machine

At the shore of the Great Salt Lake near Salt Lake City, Utah there is a candy shop that specializes in making saltwater taffy (what else?). This is a kind of sweet, chewy candy of molasses that's boiled until very thick and then pulled until the candy is glossy and holds its shape. The type of machine used to do this pulling is shown below in Figure 3.6.

Now suppose we drop two raisins into the sticky mixture being pulled by the machine, and try to follow the paths of the raisins as they are pulled to and fro by the machine. Even if the raisins start their respective journeys immediately adjacent to each other, it doesn't take too much imagination to see that they will soon be moved very far apart in the gooey mess by the stretching and folding operations of the machine. This high degree of instability in the final position of the raisins with respect to their initial positions is called *sensitivity to initial conditions,* and forms one of the most dramatic and important features of what have come to be termed *chaotic* dynamical systems.

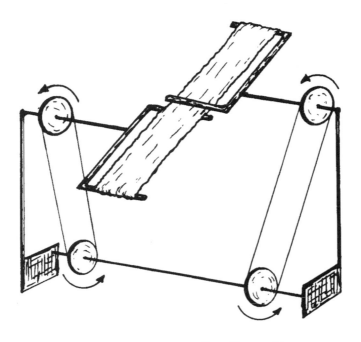

**Figure 3.6**   A saltwater-taffy-pulling machine.

As with the saltwater-taffy-pulling machine, what gives rise to this instability in chaotic systems is the combination of stretching and folding. The dynamical rule of motion for such systems moves points that begin as neighbors far apart. Because the set of all possible points for these type of systems is finite in size, things can't be stretched everywhere. This means that the set of possible points where the raisins might be located has to be folded back onto itself. This folding process then acts to bring points that are initially far apart closer together. The problem is that we cannot tell in advance which points this will be. This overall combination of stretching and folding is shown in Figure 3.7. It is this combination that leads to the instability, hence complex, unpredictable behavior of chaotic processes.

The quintessential example of this kind of instability is found in the world's weather systems. For example, Figure 3.8 shows the position of the jet stream over Europe and the mid-Atlantic in two very different weather regimes, an unsettled pattern over Britain in part (a) and a good summer/bad winter pattern in (b). The job of the meteorologist is to try to accurately predict these types of weather regimes.

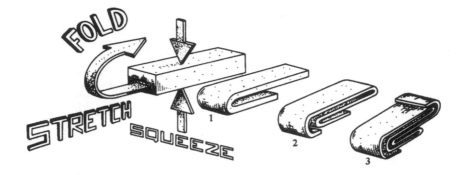

**Figure 3.7**   Stretching and folding in a chaotic dynamical system.

In an attempt to model the real atmosphere, which involves something on the order of a million variables, in the early 1960s Edward Lorenz developed a "toy" atmosphere represented by just three quantities: (a) the intensity of air movement, (b) the temperature difference between ascending and descending air currents, and (c) the temperature gradient between the top and bottom of the atmosphere. For brevity and ease of writing, let's just call these three variables $x$, $y$, and $z$, respectively. With this notation, if $x$ and $y$ are both positive or both negative, it means that warm air is rising and cold air is falling. Moreover, a positive value of $z$ signifies that the greatest temperature gradient occurs near the boundaries of the atmosphere.

In Lorenz's model, a weather state at a given moment is represented by a point in the three-dimensional abstract space whose points are determined by the values of the quantities $x$, $y$, and $z$. Thus, the development of the weather over time can be thought of as the tracing out of a curve in this three-dimensional space. The set of all possible weather states forms what is known as the *Lorenz attractor* of the

**Figure 3.8**   (a) Unsettled weather regime; (b) good summer/bad winter weather regime.

system. It's shape, shown in Figure 3.9, suggests one of the reasons why Lorenz's work on measurement sensitivity and weather forecasting is sometimes labeled the *butterfly effect*.

The first important point to note about the Lorenz attractor is that it has two separate butterfly "wings," which are abstract representations of two very different weather regimes. For the sake of discussion, let us assume that the weather dynamics in Figure 3.8 have an attractor of the type shown in Figure 3.9. Moreover, assume that the left-hand wing represents the unsettled pattern of Figure 3.8(a), while the right-hand wing stands for the pattern of fair weather in summer and bad weather in winter depicted in Figure 3.8(b).

Now consider two points that are very close to each other on the left-hand wing. These points represent nearly identical weather states in the regime characterized by unsettled conditions over the British Isles. Let's try to follow what happens to these instantaneous states of the weather as time unfolds. Figure 3.10 shows three possible histories of nearby weather states: (a) both trajectories remain on the left-hand wing, (b) both trajectories move toward the right-hand wing or (c) one

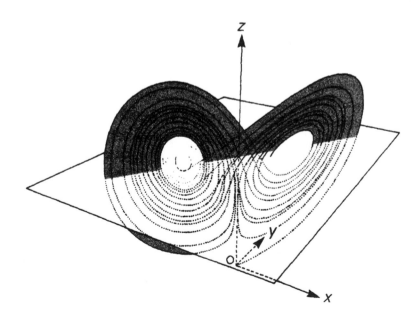

**Figure 3.9** The Lorenz attractor.

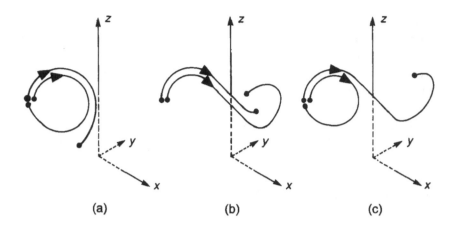

**Figure 3.10** Possible histories of two nearby weather states.

trajectory stays on the left-hand wing, while the other moves toward the right-hand pattern. It is important to note here that each of the three cases involves weather states that are initially close to each other—but each is a different initial state. The distance between the states is the same, but the initial states themselves differ in each case. It is this fact, coupled with the chaotic behavior of the dynamics governing atmospheric change, that gives rise to the different behaviors of the three resulting weather systems.

The foregoing situation shows that although the two weather states diverge in each of the three cases, thereby implying rather different forecasts of the instantaneous weather, it may still be possible to predict the overall weather regime quite far into the future. It all depends on whether we are in situations (a) and (b) or in situation (c). In the first case, the two initial weather states evolve so as to remain on the same branch of the attractor (the same wing of the butterfly), so they represent the same weather regime. But in case (c), the nearby initial states end up in entirely different regimes. Thus we conclude that although the atmosphere itself is chaotic, it may still be possible to predict overall weather patterns from certain initial states—even in the face of overall instability in the system's initial conditions.

To summarize, we have seen that complexity in a system's behavior can arise from instabilities in two very different ways. The first is when the actual system dynamics itself is changed by varying the rule of motion of the system. In this situation, there may be critical regions in the set of all possible rules of motion where a very small change from one rule to another one nearby forces the equilibrium states of the new system to move dramatically—to jump—from those of the original dynamics. Such a jump is often called a *catastrophe,* and forms part of the general issue of *structural instability.*

On the other hand, we may encounter surprising, unpredictable behavior simply because the system dynamics are sensitive to the initial conditions. This means that an infinitesimally small perturbation of the state of the system causes the system trajectory to move off on an entirely different course, one that ultimately differs greatly from the course it would have followed in the absence of the disturbance. This is the kind of instability associated with chaotic processes, and it is responsible for our inability to predict the long-term behavior of such systems as those governing the weather or the shift of political ideologies.

As noted, instability is only one of the many mechanisms at work leading to the surprising behavior of complex systems. Let's now consider another, the case when there may be no computable rule at all governing how the system operates.

## The Rules of the Game

El Farol is a sawdust-on-the-floor, downhome-type of bar-restaurant on Canyon Road in Santa Fe, at which Irish music used to be played every Thursday evening. In an earlier era, Irish economist W. Brian Arthur of the Santa Fe Institute was fond of hoisting a pint or two on weekly outings to hear these memories of his youth. But he wasn't fond of doing so in the midst of a madhouse of pushing-and-shoving drinkers and lounge lizards. So Arthur's problem each Thursday was to decide whether the crowd at El Farol would be so large that the spiritual uplift he received from the music would be outweighed by the irritation of having to listen to the performance drowned in the shouts, laughter, and raucous conversation. Being a man of logical leanings, Arthur attacked the question of whether or not to attend in analytical terms. In the process he came to some striking conclusions about that bugaboo of the economic profession, rational expectations.

Assume there are 100 people in Santa Fe, each of whom, like Arthur, would like to listen to the music. But none of them wants to go if the bar is going to be too crowded. To be specific, suppose that all 100 people know the attendance at the bar for each the past several weeks. For example, such a record might be ... 44, 78, 56, 15, 23, 67, 84, 34, 45, 76, 40, 56, 23, and 35 attendees. Each individual then independently employs some prediction procedure to estimate how many people will appear at the bar on the coming Thursday evening. Typical predictors of this sort might be:

**a.** the same number as last week (35);
**b.** a mirror image around 50 of last week's attendance (65);
**c.** a rounded-up average of the past four weeks' attendance (39);
**d.** the same as two weeks ago (23).

Suppose each person decides independently to go to the bar if his or her prediction is that fewer than 60 people will go; otherwise, the person stays home. In order to make this prediction, every person has his or her own individual set of predictors and uses the currently most accurate one to forecast the coming week's attendance at El Farol. Once each person's forecast and decision to attend has been made, people converge on the bar, and the new attendance figure is published the next day in the newspaper. At this time, everyone updates the accuracies of all of the predictors in his or her particular set, and things continue for another round. This process creates what might be termed an "ecology" of predictors.

The problem faced by each person is then to forecast the attendance as accurately as possible, knowing that the actual attendance will be determined by the forecasts others make. This immediately leads to an "I-think-you-think-they-think- ... "-type of regress—a regress of a particularly nasty sort. For suppose that someone becomes convinced that 87 people will attend. If this person assumes others are equally smart, then it's natural to assume they will also see that 87 is a good forecast. But then they all stay home, negating the accuracy of that forecast! So no shared, or common, forecast can possibly be a good one; in short, deductive logic fails. So, from a scientific point of view, the problem comes down to how to create a *theory* for how people decide whether or not to turn up at El Farol on Thursday evening and for the dynamics that these decisions induce.

## A Silicon Surrogate

It didn't take Arthur long to discover that it seems to be very difficult even to formulate a useful model of this decision process in conventional mathematical terms. So he decided to turn his computer loose on the problem and to create the "would-be" world of El Farol inside it in order to study how electronic people would act in this situation. What he wanted to do was look at how humans reason when the tools of deductive logic seem powerless to offer guidelines as to how to behave. As an economist, his interest is in self-referential problems—situations in which the forecasts made by economic agents act to *create* the world they are trying to forecast. Traditionally, economists look at such worlds using the idea of *rational expectations.* This view assumes homogeneous agents, who agree on the same forecasting model and know that others know that others know that ... they are using this forecast model. The classical view then asks which forecasting model would be consistent, on the average, with the outcome that it creates. But nobody asks how agents come up with this magical model. Moreover, if you let the agents differ in the models they use, you quickly run into a morass of technical—and conceptual—difficulties.

What Arthur's experiments showed is that if the predictors are not too simplistic, then the number of people who will attend fluctuates around an average level of 60. And, in fact, whatever threshold level Arthur chose, that level always seemed to be the long-run average of the number of attendees. In addition, the computational experiments turned up an even more intriguing pattern—at least for mathematical system theorists: The number of people going to the bar each week is a purely deterministic function of the individual predictions, which themselves are deterministic functions of the past number of attendees. This means that there is no inherently random factor dictating how many people actually turn up. Yet the outcome of the computational experiments suggests that the actual number going to hear the music in any week looks more like a random process than a deterministic function. The graph in Figure 3.11 shows a typical record of attendees for a 100-week period when the threshold level is 60.

These experimental observations lead to a fairly definite and specific mathematical conjecture:

- Under suitable conditions (to be determined) on the set of predictors, the average number of people who actually go to the bar converges to the threshold value as the number of time periods becomes large.

System theorists might also like to ponder the related conjecture:

- Under the same set of suitable conditions on the sets of predictors, the time-series of attendance levels is a deterministic random process, that is, it's "chaotic."

The major obstacle to a resolution of these conjectures is the simple fact that there currently exists no mathematical formalism within which to even meaningfully phrase the questions. I've just done it in a few words in everyday English, and Arthur created a world within which to experiment with these conjectures in a few lines of computer code. But to date there is no mathematical formalism that we can use to actually *prove* (or disprove) these conjectures.

## Inductive Rationality

The upshot of the El Farol Problem is that there exists no deductive chain of reasoning that will enable an individual to best decide whether to go to the bar. The implications of this fact for economies is clear. A forecast is only "fit," evolutionarily speaking, if it performs well in

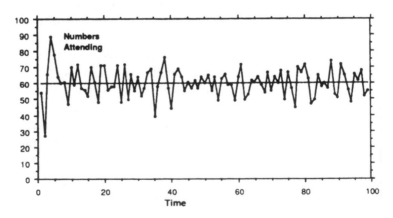

**Figure 3.11**    A simulated 100-week record of attendance at El Farol.

the world created by all the forecasts made by all the economic agents. Some predictors are mutually supporting, while some others (like cycle predictors) are mutually negating. And, as the El Farol Problem shows, common expectations are automatically negated. This diabolical twist leads to the view of an economy as a coevolutionary world, in which new predictors are constantly being created so that agents can get a temporary advantage on their cohorts. But as others discover better ideas for prediction, these new rules for action are themselves soon undone—simply by their own success. And so it goes, new predictors constantly replacing the old, with the rule of survival in the marketplace leading to the famous "Red Queen Hypothesis" of ecology: "You have to evolve as fast as you can just to stay in the game."

How do real people actually behave in situations like the El Farol Problem? The answer is what Arthur terms *inductive rationality.* Conventional economic wisdom dictates that agents can process the information available to them in a purely logical, deductive fashion to arrive at the best decision to take in any given situation. Here we see that the right thing to do—to go or not to go—depends on what everyone else is doing. But since no individual knows what everyone else is doing, all he or she can do is apply the predictor from his or her set of predictors that has worked best so far. The inductive part comes from recognizing that the right predictor to use this week may not be the one used the previous week; as new information comes in about the total attendance each week, people must reevaluate the effectiveness of their particular set of predictors, a process that may result in their changing what they regard as their "best" predictor at any given time. So agents continually shift from one predictor to another in a purely inductive fashion in an attempt to make the best possible decision about what to do. And this is exactly the way Arthur's surrogate music fans behave in his silicon world of El Farol.

Now let's swing to the other end of the spectrum and look at an example of a situation in which there are too many possible rules for action for us to ever make effective use of any of them.

## The Complexity of Economies

Neoclassical economics rests on the idea that if you have a group of economic agents, each of whom has an endowment of goods and a set of unfulfilled desires (more prosaically, supplies and demands), then there exists a set of prices for the goods at which all agents can trade what they have for what they want. In other words, at this magical price level, all markets clear, that is, there is no excess demand in the economy.

The Scottish economist Adam Smith postulated the existence of such equilibrium price levels in the eighteenth century, levels set by an "invisible hand" that balances supply against demand. But it was not until the work of people like Kenneth Arrow and Gerard Debreu nearly 200 years later that the existence of such prices was mathematically established, using techniques like the Brouwer Fixed-Point Theorem that were only invented in the early twentieth century. So now we know that such equilibrium price levels do indeed exist. If Adam Smith's invisible-hand story holds, then at least one of the price equilibria must suck-in any prices that depart from it just a little bit. In other words, it is what's called a stable *attractor* of the flow of prices as trading takes place. In summary, there must be a rule (technically, the vector field of the price dynamics) that the price movements must follow in order to reach the nirvana at which all agents in the market are satisfied.

In 1972, Hugo Sonnenschein shocked the mathematical economics community by showing that the rule of price adjustment arising from a given set of agent preferences and endowments can be literally *any* rule you like. In particular, it need not be a rule that leads to one of Adam Smith's invisible-hand equilibria. The technical details surrounding this result are not especially important for our purposes, and the interested reader can find them in the material cited in the references. What is important, however, is the fact that, in light of Sonnenschein's result, the invisible-hand theory of price movement looks a lot more invisible than even Adam Smith would have thought. In fact, it looks just plain nonexistent!

So instead of having no deductive rule by which to make decisions, as in the case of the El Farol Problem, what we have for exchange economies is an embarrassment of riches. There are just too many rules for us to ever imagine that any one of them is the rule that would be selected by a real economy. This fact takes us full circle from the El Farol situation, because if any rule might be the one governing the economy, then there really is no rule; the price dynamics can be as complicated and, thus, the economy can be as complex as one wishes.

Both the El Farol Problem and Sonnenschein's result on economic complexity show ways that complex behavior can arise as a consequence of their being no deductive rule governing the system's activity. Because a deductive rule lies at the heart of both what we mean by a mathematical proof and by a computational algorithm, let's explore this issue just a bit further to see how deeply the idea of deductive inference pervades all of mathematical and computer modeling.

## Proofs and Programs

Almost from the moment Aristotle formulated the rules of logical inference nearly 2,000 years ago, the standard by which certain knowledge is measured is the degree to which we can offer a chain of deductive arguments starting with given assumptions and ending at the statement to be proved. So, for instance, there is the classical logical argument by which we conclude with certainty that "Socrates is mortal":

| | |
|---|---|
| (Major) Assumption: | All men are mortal. |
| (Minor) Assumption: | Socrates is a man. |
| Conclusion: | Socrates is mortal. |

Mathematicians quickly translated this kind of deductive reasoning into something called a *formal system,* by which propositions about mathematical objects like numbers and geometric objects could be either established beyond doubt or definitively refuted. The basic idea underlying a formal system is that the mathematical proposition under investigation can be translated into a string of abstract symbols, just as any statement in the English language can be coded into a string of 0s and 1s by the ASCII code discussed in chapter 2. Thus, starting with a set of axioms, propositions whose truth is self-evident, one could try to use the rules of inference of the formal system to proceed from an axiom to the symbol string representing the proposition we want to prove.

For a long time, it was an axiom of faith among mathematicians that every proposition could be either proved or disproved by such a deductive procedure. One can imagine what a shock it was when, in 1931, Kurt Gödel showed that this comfortable illusion was just that—an illusion. Gödel's famous Incompleteness Theorem shows that there are arithmetical assertions, propositions about the whole numbers, that can be neither proved nor disproved using the conventional formal structures of mathematics. Such statements are simply undecidable. More recently, Gregory Chaitin of the IBM Research Division has shown that these "Gödel sentences" are not mere mathematical curiosities by displaying explicit examples of such undecidable statements—in fact, by displaying an infinite number of them.

One way of interpreting the results of Gödel and Chaitin is to say that there are propositions about numbers whose truth or falsity will never be known by following a set of rules. Such statements are forever undecidable. If we want to use their truth or falsity upon which we can build new mathematical structures, then we must take them as additional axioms; they cannot be logically derived from the original set. This means that there is no magic set of rules that will enable us to generate all the true statements about numbers, like $2 + 2 = 4$, and no false ones, such as $3 \times 3 < 0$. Arithmetic transcends rationality, in the sense that there are arithmetic statements whose truth or falsity will never be known by following a set of deductive rules.

Just five years after Gödel's work, Alan Turing produced the Turing machine model of computation discussed in the last chapter. As we saw, Turing's work proved that a simple rule-following device, the Turing machine, suffices to characterize mathematically all computing devices currently used in daily life. In the course of developing his mathematical model of computation, Turing asked a very natural question: Given a particular program $P$ for a computing machine and a set of input data $I$ that the program is to process, does there exist a procedure by which we can tell *in advance* whether or not the program $P$ will stop after a finite number of steps when processing the data $I$? Here by a *procedure* Turing meant another program that would accept $P$ and $I$ as inputs, and after a finite number of steps stop with the result of the computation written on the machine's tape. Clearly, the finiteness hypothesis is needed, since we want to actually know the result of the computation, and this will certainly not be the case if it takes an infinite number of steps to arrive at the final result. Note that what Turing was after was a *single* such program that would work for settling the halting or nonhalting of *all* possible programs $P$ and inputs $I$. To cut to the chase, Turing's famous 1936 paper established that this Halting Problem has no solution; there exists no such procedure.

Perhaps surprisingly, it turns out that Turing's theorem on the unsolvability of the Halting Problem and Gödel's Incompleteness Theorem are exactly the same result, one expressed in the setting of formal logical systems, the other in the context of computing machines and programs. This equivalence is perhaps easiest to see simply by putting the two results side by side:

---

**GÖDEL'S INCOMPLETENESS THEOREM** *For any consistent formal system $\mathcal{F}$ purporting to settle, that is, prove or disprove, all statements of arithmetic, there exists an arithmetical proposition that can be neither proved nor disproved in this system. Therefore, the formal system $\mathcal{F}$ is incomplete.* ∎

---

**THE HALTING THEOREM** *For any Turing-machine program $\mathcal{H}$ purporting to settle the halting or nonhalting of all Turing-machine programs, there exists a program $\mathcal{P}$ and input data $\mathcal{I}$ such that the program $\mathcal{H}$ cannot determine whether $\mathcal{P}$ will halt when processing the data $\mathcal{I}$.* ∎

These results tell us about the existence of events in the mathematical realm that are not the end result of following a set of rules, a program. Whether analogous events occur in the worlds of natural and human affairs is an entirely different matter, but the problem of protein folding, discussed in the last chapter, as well as the $N$-Body Problem of celestial mechanics, considered earlier, give us considerable food for thought. However, it should be kept in mind that such impossibilities are mathematical consequences of a particular *model* of computation, the model based upon the Turing machine. Recently, people have begun to consider other models that may help us break the Turing-machine barrier. Perhaps the most exciting such model is a biological one, in which the folding of cellular DNA is used to calculate the incalculable.

## DNA as a Super-Turing Machine

In 1994, the calculating power of billions of molecules was tapped for the first time in a revolutionary new type of DNA computer, when Leonard

Adleman of the University of Southern California encoded the input details of a computational puzzle into single DNA strands. He then mixed these strands together, allowing them to link up into many different double helix output molecules, among which a few had a structure that encoded the answer to the puzzle.

The problem Adleman looked at involved finding a specific path through a network of points. An example might involve the four cities New York, Paris, Vienna, and Tokyo. Nonstop flights are scheduled only from New York to Paris, Paris to Vienna, Paris to Tokyo and Vienna to Tokyo. A question one might ask is: By which route from New York to Tokyo can a traveler visit all the cities, taking only three journeys?

In this case the answer is obvious: Fly from New York to Paris, then to Vienna, then on to Tokyo. But if the problem involved all the major cities in the world, and all the connecting flights, the number of possible itineraries would become astronomical. On such a large scale, this type of problem is impossible to solve even with the fastest supercomputers, because the time taken to find the result grows exponentially with the number of destinations.

Adleman solved a seven-city problem of this type by encoding the details into single strands of DNA. The double helix of DNA is formed of two complementary strands of the four DNA bases, labeled A, T, G, and C. The base A is the complement of T—that is, it can bind only to T. Similarly, G is the complement of C. One strand of the DNA molecule is thus a complementary mirror image of the other. Adleman randomly selected single strand codes to represent each city—say, ATGCGA for New York, TGATCC for Paris, GCTTAG for Vienna. Then the strand representing each flight path might be defined by the last three code letters of the city the flight was leaving, and the first three code letters of the destination. So a flight New York-to-Paris would be coded CGATGA in the earlier four-city example.

Using genetic engineering, it is possible to manufacture single DNA strands to order. Adleman mixed the complement strands of the city DNA (ACTAGG for Paris, say) with the flight path strands in the test tube, and, as they joined to form double helices, the flight path strings acted as complementary bridges to bind the DNA city strings together. Molecules for all possible combinations of flights formed, but given that billions of molecules were reacting, it was almost certain that a molecule representing the correct flight path would be in the final mixture.

**Plate I**   (see Figure 1.1, page 5)

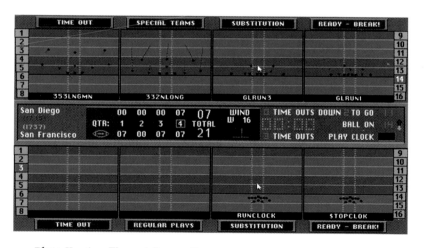

**Plate II**   (see Figure 1.2, page 5)

**Plate III**  (see Figure 2.6, page 44)

**Plate IV**   (see Figure 2.13, page 69)

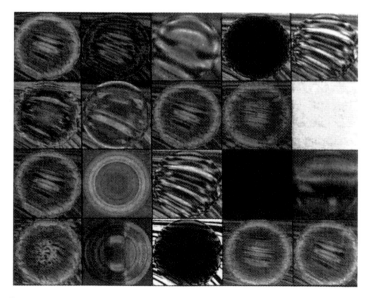

**Plate V**   (see Figure 2.14, page 70)

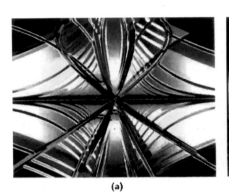

**(a)**

```
(round(log(+y (color-grad(round(+abs (round
(log(+y(color-grad(round(+y(log(invert y) 15.5))
x)3.1 1.86#(0.95 0.7 0.59) 1.35))0.19)x))(log
(invert y)15.5))x)3.1 1.9#(0.95(0.7 0.35)1.35))
0.19)x)
```

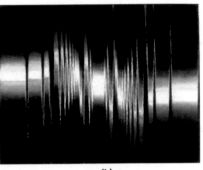

**(b)**

```
(rotate-vector(log(+y(color-grad(round(+(abs
(round(log #(0.01 0.67 0.86)0.19) x))(hsv-to-
rgb(bump(if x 10.7 y)#(0.94 0.01 0.4)0.78#(
0.18 0.28 0.58)#(0.4 0.92 0.58)10.6 0.23
0.91)))x)3.1 1.93#(0.95 0.7 0.35)3.03)) 0.03)
x#(0.76 0.08 0.24))
```

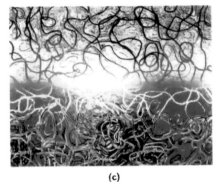

**(c)**

**Plate VI**　　(see Figure 2.15, page 71)

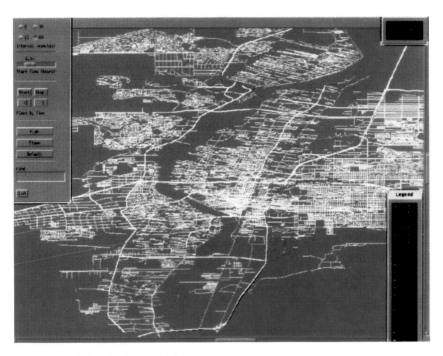

**Plate VII** (see Figure 4.5, page 138)

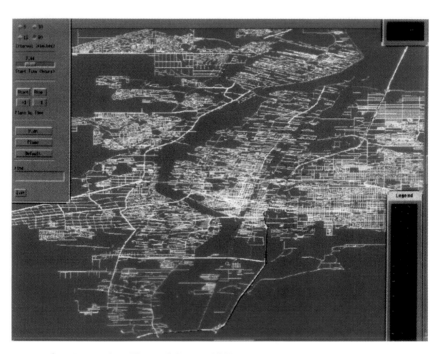

**Plate VIII**   (see Figure 4.6, page 139)

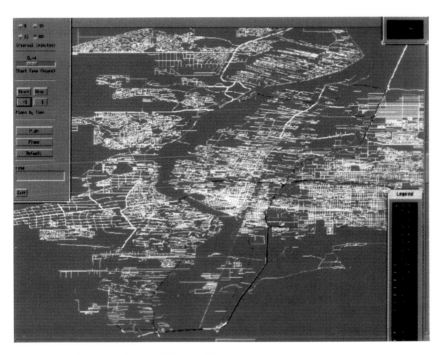

**Plate IX**    (see Figure 4.7, page 140)

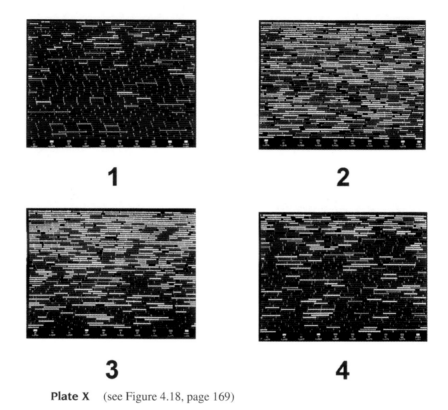

**1**

**2**

**3**

**4**

**Plate X**    (see Figure 4.18, page 169)

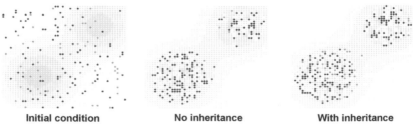

Initial condition          No inheritance          With inheritance

**Plate XI**    (see Figure 4.19, page 172)

The problem was to identify the output molecule. It has the property that it spans the length of all the city codes, beginning with the starting city and ending with the destination, and it contains the codes of all other cities en route only once. Knowing this, Adleman was able to isolate the molecule representing the solution using standard techniques of molecular biology. He then analyzed the order in which its building blocks had been put together, revealing the correct sequence of cities.

One exciting aspect of the DNA computer is its potentially unprecedented energy efficiency and minimal memory requirements. Storing information in DNA requires about one-trillionth the space required by existing storage media, such as magnetic tape. However, in its present form this method of computing is not practical enough to replace conventional computers; the solution to the seven-city problem emerged only after seven *days* of work rather than the few picoseconds the problem would have taken on a conventional supercomputer. If future research can find shortcuts, biological computers may become a practical method for solving currently intractable network-type calculations often encountered in real life, such as the design of telephone-routing systems. Even more interesting from a theoretical point of view is that such computers may allow us to actually calculate "uncomputable" quantities that are beyond the reach of machines based on the Turing model of computation.

With these ideas of computation in hand, it's easier to see why unanticipated behavior might manifest itself in computer simulations as a result of the existence of real-world phenomena that cannot be captured completely in a finite set of rules. We discussed this earlier in connection with things like neuronal firing in the brain. But the *possibility* of such uncomputable processes has much broader currency. In fact, it is the purpose of some simulations to explore the very existence of such phenomena outside mathematics. Much more about this will be covered in the final chapter. For now, let's turn our attention to how connective structure among a system's components can act as a generator of surprise.

## The Connections That Count

Every cell in the human body contains approximately 100,000 genes—including an unknown number of regulatory genes—all switching each other on and off in an unimaginably complicated network of interactions. Stuart Kauffman, a theoretical biologist at the Santa Fe Institute, has spent the last 30 years trying to explain the puzzling fact that all this switching on and off doesn't lead to utter chaos, but rather results in the cell organizing itself into stable patterns of activity appropriate for its particular function in the body. How is it that this seemingly random operation of individual genes leads the cell to configure itself into a stable, workable structure? Speaking solely in Darwinian terms, it is difficult to understand how new types of organisms could possibly arise out of merely random mutations and natural selection—the standard Darwinian party line. Something more seems to be needed to account for the great diversity of living forms surrounding us today. The answer, according to Kauffman, lies in the marked preference of complex systems to spontaneously organize themselves into persistent patterns of activity that work. As Kauffman puts it, "Darwin didn't know about self-organization."

In contrast to mainstream biologists and chemists, who try to explain the emergence of new patterns by looking at genetic regulation in painstaking biochemical detail, Kauffman has built a mathematical model in the form of a network of interactions that mimic the genetic regulatory activity. Suppose we have a network of $N$ genes, each regulated by $K$ other genes, each of which can be either ON or OFF at any given moment. Thus, there are a total of $2^K$ possible inputs to each gene in the network. At each moment, Kauffman assumes that one of these input patterns is selected at random. Moreover, he also chooses one of the $2^K$ possible ON/OFF patterns for each set of inputs to each gene. This pattern at each gene determines whether the gene will be ON or OFF at the next instant. For these *Kauffman networks,* the rule of state transition that turns genes ON or OFF is chosen randomly at each moment, as is the neighborhood of each cell, which may be decidedly nonlocal.

A simple example of such a network is shown in Figure 3.12, in which all the elements are ruled by the Boolean OR function. This means that the gene is ON at the next time period if *any* of its inputs is ON at the current period. Initially, all the elements are OFF (time 1). Changes cascade through the system after one gene turns ON (time 2). Because of the way the network is connected, some genes end up "freezing" into the ON state (time 6). Moreover, these elements will return to the ON state even if they or one of their inputs is altered.

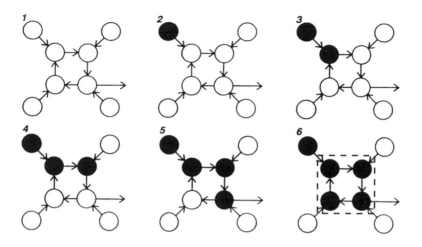

**Figure 3.12** A simple Kauffman net.

In numerous experiments with different values of $N$ and $K$, what Kauffman saw was not total chaos at all. Rather, such networks display a powerful tendency toward self-organization by settling into a small number of different long-run behaviors that are periodic. In other words, Kauffman saw sets of configurations that simply repeated themselves over and over indefinitely. For example, when $K = 2$, the length of these cycles, as well as the total number of different cycles, is small, typically on the order of $\sqrt{N}$ in both cases. Kauffman believes these cycles can be identified with the possible cell types (liver, stomach, skin) that may arise from the genetic network. If he's right, then a cell type is a stable recurrent pattern of gene expression, created solely by the logical connective structure built into the genetic network.

In Kauffman networks, gene $A$ turns on gene $B$, which then turns on gene $C$ and turns off gene $D$ and so on. What this type of model shows is that stable cellular types may arise spontaneously as the attractors of a dynamical system. It is through the subtle interplay between the stable and unstable attractors—cooperation and competition—that patterns of change and periods of stasis can slowly evolve. Furthermore, Kauffman's models tell us that the formation of these stable patterns is almost inevitable, regardless of how disorganized the network is to begin with. Basically, the genetic interaction dynamics seem to force the cellular genome to spontaneously organize itself into a structure that can survive in its environment.

Kauffman networks portray the connections between elements in a Boolean network as entities that are either present or absent, regardless of how many genes serve as input to a given gene. But it stands to reason that if a particular gene has many more inputs than another, this fact is significant for the functioning of the gene in the overall network; hence, the dimension of the connection (the number of input genes) is important in understanding how signals pass from gene to gene through the net. This basic idea has been exploited by mathematicians Ron Atkin and Jeff Johnson to develop a theory of connectivity termed *q-analysis,* for the study of connective patterns in systems ranging from road traffic networks to the structure of modern art. Such dimensional connections enable us to see deeper into the structure of the game of chess.

### Positional Play in Chess

Emmanuel Lasker reigned as the world chess champion for 27 years. Unlike many chess geniuses, Lasker's interests were far from narrow, and his concern with philosophical matters led to a deep consideration of what he called the "philosophy of struggle." For Lasker, the chessboard was a stage reflecting the struggle of life in its purest form, a view encapsulated in his well-known remark, "On the chessboard lies and hypocrisy do not long survive." He went on to note that "there are sixty-four squares on the chessboard; if you control thirty-three of them you must have an advantage." Although this is a vast oversimplification of the situation, it points out the importance of positional play in the thinking of chess masters. This positional, or strategic, view of the game also suggests that it should be possible to use the multidimensional perspective of $q$-analysis to evaluate board positions. Let's see how this might be done.

The game of chess can be considered as a relation between the squares of the board and the Black and White pieces. Actually, there are at least two important relations here: (a) a relation $R_B$ between the Black pieces and the squares and (b) a relation $R_W$ linking the White pieces and the squares. These relations might take many forms. For instance, we could define a relation by saying that a White piece and a particular square are related if that piece occupies the given square. However, such a relation, although perfectly well defined, is completely useless because it does not embody any of the rules of the game of chess.

In order to incorporate the rules of the game into some meaningful relations $R_W$ and $R_B$, let's first define what we mean by a man (pawn, Knight, Bishop, Rook, Queen, or King) *attacking* a given square. (Note: Here we shall adopt standard chess jargon, calling all the chessmen *men* while reserving the term *pieces* for those men that are not pawns.) For the sake of definiteness, let us center attention for the moment on the White men. We say that the White man $P$ *attacks* square $S$ if exactly *one* of the following conditions holds:

1. If it is White's move and $P$ is not a pawn or the White King, then $P \rightarrow S$ is a legal move.
2. If $P$ is a pawn, then $S$ is a capturing square for $P$ (this means that $S$ is one of the two squares diagonally in front of $P$).
3. If there is a White man $Y$ on square $S$, then $P$ is protecting $Y$ (in the ordinary chess-playing sense of one man protecting another).
4. If $P$ is the White King, then $S$ is adjacent (horizontally, vertically, or diagonally) to the square occupied by $P$.
5. If it is White's move and square $S$ contains a Black man $Z$ (other than the King), then $P$ captures $Z$ is a legal move. (Note: If $Z$ is the Black King, we are at checkmate and the game is over.)
6. The Black King is on square $S$ and is in check to $P$.

With this idea of *attacking* in mind, we then define the relation $R_W$ linking the White men and the squares of the board as follows:

"White man $P_i$ is $R_W$-related to square $S_j$ if and only if $P_i$ is *attacking* square $S_j$."

The corresponding relation $R_B$ for the Black men is defined analogously. Note that in these relations a piece does not attack its own square. This means that the pieces and pawns must protect each other by attacking the occupied squares. Now let's see how these relations can be used to understand the relative strengths of the pieces and to examine how they combine their varied abilities to form a "field of force" at any particular stage of play.

Let us agree to call a particular distribution of the White and Black men on the board a *mode.* The game begins with White to move in mode [0, 0]. After White makes the first move, the game enters mode [1, 0]; after Black's first move the game is in mode [1, 1] and

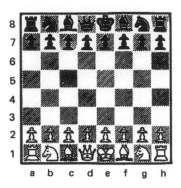

**Figure 3.13** Mode [0, 0] at the beginning of a chess game, White to move.

so on. Figure 3.13 shows the distribution of the playing pieces when the game is to begin in mode [0, 0]. Consider the relation $R_W$, expressing the way the White men are linked by attacking squares on the board.

By thinking of the way in which the black and white pieces attack squares on the board as a connective structure linking the squares and pieces, we gain a new perspective on the traditional arguments for the relative strengths of the playing pieces. Taking the value of a pawn to be P = 1, the traditional values assigned to the pieces are B = N = 3, R = 5, Q = 9, K = infinite. Leaving the King out of consideration, we can reasonably measure the relative strengths of the pieces in any playing mode by their "dimension," which is simply the number of squares the piece attacks minus 1. At the outset, when the game is in mode [0, 0], these values are Q = 4, KN = QN = 2, with all the rest of the men having dimension 1 except for the KRP and the QRP which have dimension 0. These values shift during the course of the game. For example, if White makes the King's pawn opening move e2 → e4, the game enters mode [1, 0], and it's easy to see that the dimensions of all the White men remain the same except for the Q and KB. In this new mode, the dimension of the Queen is now 7 instead of 4, whereas the KB has increased its dimension from 1 to 5.

Following the above line of argument, it's instructive to consider what the *maximum* possible dimensions of the pieces can be in the absence of obstacles on the board. R. H. Atkin has computed these quantities to be those shown in Table 3.2.

Comparing these maximum dimensions with the traditional values given earlier, we find pretty good agreement, with the exception of the

**Table 3.2**  Maximum dimensions of the chess pieces.

| Piece | Maximum Dimension | Maximum Attained |
|---|---|---|
| Pawn | 1 | in all modes |
| Knight | 7 | when N is inside the square c6–f6–f3–c3 |
| Bishop | 12 | when B is on square d5, e5, d4, or e4 |
| Rook | 13 | when R is on any square |
| Queen | 26 | when Q is on square d5, e5, d4, or e4 |

greatly increased value of the Bishop as compared with the Knight. We notice also the great importance of the central squares d5, e5, d4, and e4. Various theories of chess openings have emphasized the significance of controlling these squares. The dimensional analysis also enables us to appreciate schools of thought devoted to the argument that opening play should be dedicated to the task of bringing the Rook into play as soon as possible. Castling, as well as the classical King's Gambit opening with its early sacrifice of the KBP, are devices for accomplishing just this. This kind of analysis enables us to attach some geometrical structure to any particular mode of the game. For more details on this type of analysis, we urge the reader to consult the works by Atkin cited in the references.

Both Kauffman nets and $q$-analysis show us how the connective structure of a system can lead to interesting, hidden patterns in either its static or dynamic behavior. These patterns, in turn, generate surprise. As our final stop on this tour of surprise-generating mechanisms, we now look at what is probably the most important mechanism of all, the phenomenon of emergence.

## The Laws of Emergence

During the 1994–1995 National Football League season, a total of 224 regular-season games were played. Suppose you decide to bet $10 on each game with a friend, who kindly allows you to pick the winner of each game by simply tossing a coin. Since you do not have an infinite bankroll, a quantity of some concern to you in such an experiment is $T_n$, your total take after the first $n$ games have been played. The quantity $T_n$ is a random variable, whose actual value depends on the outcome of each of the $n$ games. These outcomes, in turn, are determined by myriad independent random factors, such as a defensive back slipping on a wet piece of turf at a crucial moment or the wind happening to gust in a particular direction just at the moment of a crucial field goal attempt. Such factors can be neither quantified nor fully understood, so the actual result of any particular game is itself a random variable. At first glance, it appears the prospects of being able to say anything mathematically meaningful about the way the possible values of $T_n$ might distribute themselves are rather bleak, because the specific value that $T_n$ assumes for any sequence of $n$ games depends on the way these manifold random factors happen to come out—the way the ball bounces, so to speak. But in mathematics, intuition is as much an enemy as a friend.

Following earlier work by Gauss and others on the distribution of random errors, in 1887 the Russian mathematician Pafnuty L. Chebyshev succeeded in proving that under very weak assumptions about the properties of the random factors determining the specific value of $T_n$, the possible values of $T_n$ will *always* distribute themselves in accordance with the famous bell-shaped curve of the normal probability distribution. This is the content of the celebrated Central Limit Theorem (CLT) of probability theory, one of the most remarkable—and useful— mathematical results of all time. More specifically, the CLT says that if the individual random factors are numerous, mutually independent, and the effect of each such factor on the total payoff is very slight, then $T_n$ will be a normally distributed random variable. So for a reasonably large sample of trials, such as the $n = 224$ games played during the course of an NFL season, you can be very confident that not only will your expected total payoff from betting on each game be 0, but that the deviation from this amount will vary in accordance with the dictates of the normal distribution, basically ensuring that with 95 percent confidence you can expect to win or lose no more than about $220.

Less well chronicled, but of even greater interest for system theorists and "complexicists," is that the CLT serves as a prototype for what we might call a law of complex systems. If having a theory of complex systems means anything, it means having a set of laws, or principles, by which systems organize themselves and behave. For example, a useful theory of complex systems would offer laws explaining/predicting why political ideologies peak and decline, how immune systems adapt to threats, and when price bubbles and crashes will occur in the stock market. Moreover, such a theory of complex behavior will offer the *same* laws underlying each of these very different types of processes. The CLT does precisely this in a very specific and definite way by considering a large number of independent agents, interacting in such a fashion that no one agent can exert very much influence on the system behavior as a whole. Beyond this, the theorem imposes no conditions on how the agents behave. Yet out of such lowly assumptions comes the amazing fact that a large collection of such interacting agents must necessarily generate aggregate behavior that is normally distributed. When phrased this way, it's easy to see how one can claim that the CLT is the forerunner of many mathematical accounts of emergent behavior.

The CLT is a law about how simple behavior emerges out of the interaction of a large number of complex agents, but emergence can go in the other direction, as well. For example, human societies, ant colonies, and cellular slime molds all consist of large numbers of individual agents that are relatively simple in their behavior. Yet, when these agents interact, again by rather straightforward rules, the resulting social structure turns out to be a rich and varied one, having a level of complexity far greater than any one of its components.

By now, more than a century after Chebyshev's work, most people probably regard the CLT as being about as airtight as a mathematical fact can be. However, the CLT did not always rest on such solid ground, and for a long time it was clouded in mystery deep enough to prompt the famed French mathematician Henri Poincaré to once remark that, "mathematicians regard it as a physical law, whereas physicists hold mathematicians responsible for it."

The history of the CLT up to its proof as a formal mathematical fact is very reminiscent of the position occupied by a number of the empirical relations that are regularly offered up today as candidates for laws of complex behavior. So we might well ask if it is possible that there is something to be learned about the development of a theory of complex systems by examining the life and times of the CLT? More specifically, could we possibly learn about how an empirical relation arising in the study of different types of complex systems could come to eventually be regarded as a law of nature?

## Power Laws and Languages

One type of empirical relation that seems to turn up regularly in complex-systems studies is a *power law.* A particularly intriguing illustration of this kind of law is the relation observed between the rank order of words in a language and the frequency of the appearance of these words in a sufficiently large body of text. This relationship, now termed *Zipf's Law,* was first presented by George Zipf in his 1949 volume *Human Behavior and the Principle of Least Effort.* To describe this law, suppose we list the words of the English language according to how common they are. So, for example, the most common word is *the,* which we assign rank 1. The next most common word is *of,* which is then given rank 2, followed by *and* having rank 3, *to* with rank 4, and so on. What Zipf discovered was that if we plot word frequency in a large body of English text versus the word rank, we obtain a graph of the type shown in Figure 3.14.

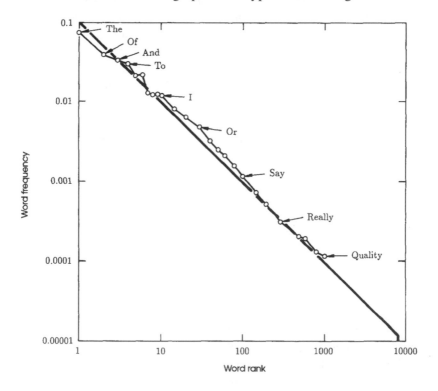

**Figure 3.14**   Zipf's Law for the English language.

Algebraically, Zipf found a relationship between word frequency and word rank: The word of rank $n$ in the language appears with frequency proportional to $1/n$. It's not important for us to go into the details of this relationship here. We need only note that this kind of relationship implies that for a good writer with an active vocabulary of, say, 100,000 words, the 10 highest ranking words occupy 24 percent of a text, whereas for a popular novel or newspaper using a slimmed-down vocabulary of 10,000 words, this percentage increases marginally to nearly 30 percent. To illustrate an in-between case, Mr. William Bowers of Thousand Oaks, California kindly sent me a Zipfian analysis of Conan Doyle's classic Sherlock Holmes story "The Hound of the Baskervilles." Using the same vocabulary size of 10,000 words, the result of Bowers's analysis of this 59,498 word story is shown in Figure 3.15. The striking conformance of the Conan Doyle story with Zipf's empirically derived relation is clear, although it is interesting to note that in this story the word *to,* which holds rank 4 in general English text, has moved down to rank 5 for this particular text.

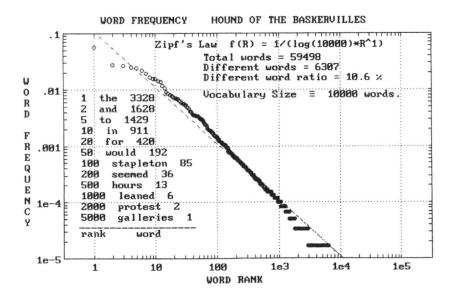

**Figure 3.15** Zipf's Law and "The Hound of the Baskervilles."

Zipf tried to derive the form of his law by appealing to a principle of least effort. In fact, it is possible to show, using information-theoretic arguments, that by some process of evolutionary selection, natural languages that survived are exactly those that were able to convey the maximum amount of information at a given cost, measured by the average time needed to produce the words of the language. Only those languages obeying the Zipf's form have this property. But wait! It turns out that monkeys hitting typewriter keys at random will also produce a "language" obeying Zipf's Law.

Suppose we consider a nine-letter alphabet with a space character, so that our mythical monkey strikes each character with likelihood 1/10. Many years ago, Benoit Mandelbrot showed that Zipf's Law for this monkey language is hardly any different than it is for English. But the median word rank in this language is 1,895,761, which means it takes this many of the most frequent words in the monkey language to reach a total probability of 1/2. By way of contrast, the analogous figure in English is between 100 words (for typical media texts) to 500 (for belle-lettres writers). Thus, the monkey language is a very wordy one compared to English.

Somewhere in between the language of English and that of the monkey is the language of DNA. Recently, physicist Eugene Stanley, in collaboration with a group of geneticists, applied the Zipf test to the part of yeast DNA that does not correspond to the coding region for any genes, the so-called junk DNA. When the researchers arbitrarily divided up the junk into "words" between three and eight bases long, what emerged was a surprisingly tight fit to the theoretical Zipf curve for natural languages like English or Chinese, as is indicated in Figure 3.16. Moreover, the researchers applied a second test to quantify the redundancy in the yeast language, finding a level significantly higher than what would be expected if the junk were completely random. Although controversial, these two findings together suggest that something is written in these mysterious regions, and that the junk may not be so much junk, after all. As Harvard biologist Walter Gilbert described it, "I think the junk is like the stuff in a junk shop. You can find lovely things in it."

Zipf's Law, be it a law of nature or an empirical relationship, shows that there is some kind of pattern lurking below the surface in the arrangement of symbols used by nature and humans to communicate information. This is not a pattern that can be seen in the individual

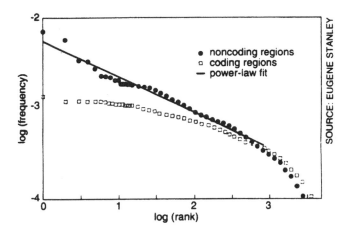

**Figure 3.16**  Zipf plot for arbitrary words in noncoding part of yeast DNA.

words of a language itself, but rather emerges from the interaction of the words to form sentences and even more complete expressions of thought. So the Zipfian-type relationship is due more to the *interaction* of the words than to the meaning of the individual entities themselves. This is the essence of what is meant by an emergent property, in this case an emergent property of languages. Many more of these kinds of properties make their appearance in the worlds discussed in the next chapter.

# CHAPTER

# 4

# Artificial Worlds

## A Louisiana Saturday Afternoon

Louisiana Boulevard in Albuquerque, New Mexico, is a fairly run-of-the-mill major surface street in your average, flat, dusty, southwestern American town. Running north and south, this four-lane thoroughfare joins two of the major shopping malls in Albuquerque before it eventually peters out in the residential neighborhoods in the foothills of the Sandia Mountains that bound the city to the east. Most of the year it's difficult to distinguish Louisiana Boulevard from other major arteries like Menaul, Wyoming, Juan Tabo, and Candelaria Boulevards, which form the Albuquerque road-traffic network. However, on Saturday afternoons between Thanksgiving and Christmas, Louisiana Boulevard's unique position as a link between the two shopping malls turns the street into a wall-to-wall parking lot, as shoppers frantically rush from one mall to another in hopeless attempts to complete their Christmas lists before the appearance of Santa, together with Rudolph and all the other reindeers.

The Christmas-shopping congestion on Louisiana Boulevard is an entirely unremarkable example of a problem that occurs every day of the year in road-traffic networks around the world. Traffic engineers have spent countless hours trying to divine the right combination of traffic-light patterns, one-way streets, left-turn lanes, freeway interchanges, bridge approaches, and the like to reduce such congestion to enable travelers to get from one place to another in a reasonable amount of time. The problem is that there is no decent theory of how to combine these various factors in any given situation to create a network that solves the problem. One of the main reasons for the absence of such a theory is that there is no laboratory in which various hypotheses about road-traffic flow can be tested in a scientific, that is, repeatable and controllable, manner. Enter the Los Alamos National Laboratory.

A few years ago, the U.S. Environmental Protection Agency set forth regulations (the Clean Air Act of 1990) specifying environmental impact standards for just about any change that anyone might want to make to anything that involves the air, earth, fire, and water constituting the human habitat. In particular, these standards apply to proposed modifications to road traffic systems, changes such as the construction of high-speed rail links, the addition of a bridge, or the construction of a freeway. Unfortunately, there is no known way of actually assessing whether any proposed change of this sort actually meets the standards laid down by the Clean Air Act of 1990. In 1991, Los Alamos researcher Chris Barrett had the bright idea that computing technology had finally reached a level at which it should actually be feasible to build an electronic counterpart of a city like Albuquerque, complete with every individual street, house, car, and traveler. Barrett thought that with such a surrogate version of the city inside his computer, it would then be possible to couple this silicon city to an air-pollution model so as to actually calculate directly the environmental impact of any proposed change to the road traffic system. Happily, some visionary thinkers at the Federal Highway Administration of the U.S. Department of Transportation agreed with Barrett, and provided the financial support needed to turn Barrett's fantasy into reality.

At first glance, one might think that even with the state-of-the-art computing capacity available at a place like Los Alamos, it would be an impossible task to create an "Albuquerquia" of 200,000 or so households and more than 400,000 daily travelers moving about on 30,000 road segments. Not only does this represent an imposing database manipulation problem, but there is the further task of planning travel routes and keeping track of each of these thousands of travelers every second or so as they make their way through the network. But Chris Barrett is not the type of man to be discouraged by such minor obstacles, and contrary to all expectations he and his group succeeded in creating just such a would-be world. It is called *TRANSIMS*. The overall structure of this electronic world is shown in Figure 4.1. It consists of four basic components.

**Travel Demand and Transport System Data.** This module contains the geographic structure of the city, the roads in the traffic network, households, types of vehicles, locations where people in the households work, shop, study, and play, and when they want to access these various locations. In addition, this component of *TRANSIMS* contains information about other factors such as traffic light patterns and the layout of one-way streets.

**Trip Route Plan Generation.** Once a particular traveler's origin and destination have been given, this component of *TRANSIMS* uses a fairly sophisticated sub-optimization procedure to compute a good route for getting the traveler from his or her starting point to the destination. This plan involves balancing the traveler's various preferences, such as not wanting to travel on dirt roads, preferring certain major surface streets, and minimizing travel time.

**Traffic Microsimulation.** After the travelers and their plans have all been sprinkled into the network, this module moves the travelers throughout the system. At each time step, typically 1 second of real time, the microsimulation monitors each traveler and either moves him or her along the existing plan for one more step or calculates a new plan, based upon the current information available to the traveler, factoring in things like local congestion and accidents.

**Environmental Simulation.** Each vehicle in the system is coded for its operating characteristics—properties such as engine type, exhaust system, tire pressure, and speed. As the vehicles make their way through the traffic network, these properties give rise to exhaust emissions, most importantly carbon monoxide, nitrous oxide, and sulfur dioxide. These gases are then processed by the environmental module of *TRANSIMS*

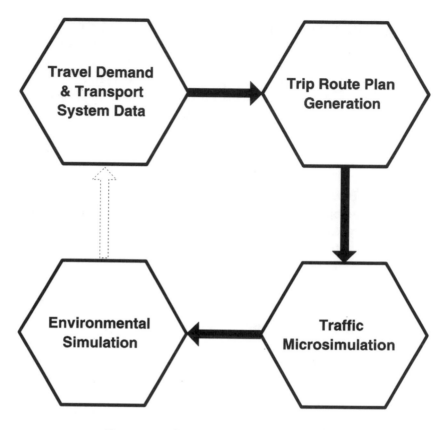

**Figure 4.1** Structure of the *TRANSIMS* world.

to produce a pattern of different gases hanging over the city. It is this part of the system that finally answers the questions the Environmental Protection Agency cares about, namely, How does a given proposed change in the system create traffic patterns and how do these patterns impact the environment?

Note that in Figure 4.1 there is a dotted arrow closing the loop between travel demand and the environment. Although this linkage does not exist as a formal structure in the *TRANSIMS* model, it is nonetheless a very real part of the system, because it represents the ways in which the physical network and the desires of travelers meet the requirements of a safe, clean overall environment. Now let's look at *TRANSIMS* in action.

### "Albuquerquia"

Albuquerque is a city of about a half million people. Geographically, it sits on a high desert plain, with the Sandia Mountains towering over the city to the east and the Rio Grande River running through the middle of town. The town's road traffic network is distinguished by two freeways, Interstate 25 and Interstate 40, that intersect in the center of the city. Interstate 25 runs north–south, while Interstate 40 crosses it in the east–west direction. Figure 4.2 gives a bird's-eye view of the city showing this topography and the two freeways. In this image, the river runs to the left of and approximately parallel to Interstate 25. Zooming in on this view, in Figure 4.3 we see some of the structure of the major surface streets in the Albuquerque road-traffic network.

As travelers are dropped into the network and begin making their way from one place to another, *TRANSIMS* enables us to take a god-like view of the system, zooming in wherever we wish to look at local traffic behaviors. For instance, Figure 4.4 (page 137) shows traffic moving from left to right along one side of Interstate 25. (Note: The traffic on the other side is not shown, because it does not interact with the vehicles that

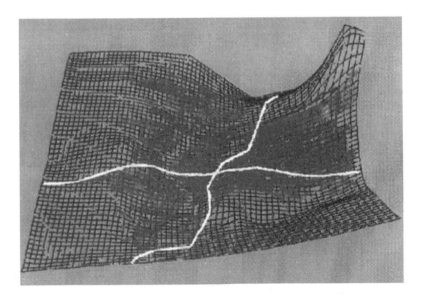

**Figure 4.2**   An overview of the city of Albuquerque, New Mexico.

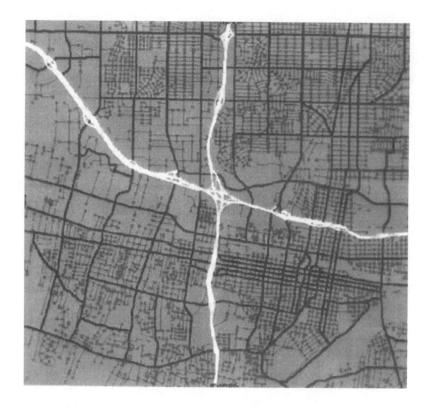

**Figure 4.3**   Some of the principal surface streets in Albuquerque.

are depicted. Thus, it would needlessly consume computer resources and slow down the system to display it.) *TRANSIMS* also enables us to look at individual vehicles in the system by freezing them at a particular moment in time. The window in the figure gives information about vehicle number 11150, which happens to be the third car from the left in the inside lane. Here we see information about that car's position, speed, acceleration, status (passing, in this case), and fuel enrichment state (for the air-pollution model). It is not without interest to note that at this moment car 11150 is traveling at a constant speed (zero acceleration) of nearly 97 mph! Clearly, there are no police patrolling the freeways of Albuquerquia (yet)!

As an illustration of the way *TRANSIMS* can be used to study traffic patterns, consider the buildup of morning rush-hour traffic. Figure 4.5 (page 138) shows the traffic density at 6:44 A.M. The densities are color-coded with the lowest density being white, moving up through medium

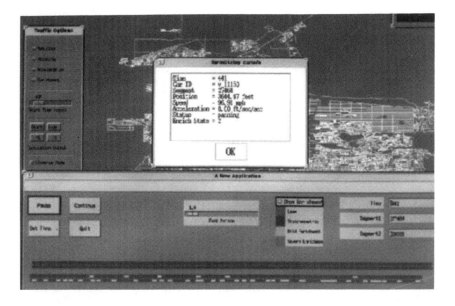

**Figure 4.4**  Freeway traffic.

densities in green and orange, to the highest densities of traffic shown in red and purple. As indicated in Figure 4.5, the early-morning traffic buildup starts on the north-south freeway, Interstate 25, on the surface streets in the east, and on the northernmost bridge across the Rio Grande. For the most part, this traffic represents movement from the residential suburbs in the south, east, and north toward the central business districts located near the center of the figure. Even at this early hour the freeway and a couple of the major surface streets are already at perilously high traffic densities in the orange and red zones.

One hour later, at 7:44 A.M., Figure 4.6 (page 139) shows that the traffic is really picking up. Now all of Interstate 25 is in the orange and red zones, as is the east–west freeway, Interstate 40, and the northern bridge across the river. Moreover, many of the east-west surface streets are now moving into the medium-density green zone, along with two of the other bridges across the Rio Grande.

Finally, at the height of the rush hour at 8:44 A.M., the system is completely clogged, as seen in Figure 4.7 (page 140). Now both freeways, all bridges, and a number of the surface streets are in the high-density orange, red, and purple zones. In addition, even many of the secondary streets in the close-in residential neighborhoods ringing the

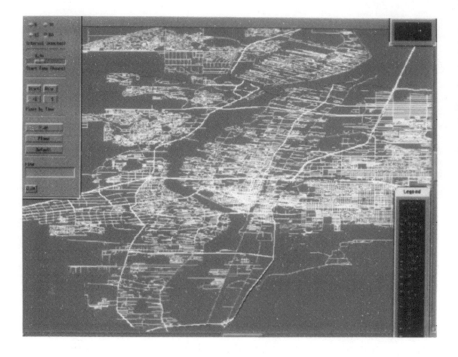

**Figure 4.5**   Traffic densities at 6:44 A.M. (see Plate VII).

central business district have moved into the medium-density green zone. At this point fully developed rush-hour congestion is in place: Drivers are drumming their fingers on the steering wheel, stuck in traffic moving slower than walking pace, and, in general, the Albuquerque system, as a road transport system, has failed in its mission of delivering people from one part of the network to another.

These results suggest countless what-if games that can be played with *TRANSIMS:*

- What effect would a new bridge across the Rio Grande have on rush-hour traffic?

- How would the traffic density on Menaul Boulevard (one of the principal east–west surface streets) change if the major east–west thoroughfares were changed to one-way flow, east to west, from 6 A.M. to 9 A.M.?

- What if traffic were metered entering Interstate 25 and/or Interstate 40?

- How do the traffic-light patterns on Central Avenue (another major east–west artery) affect rush-hour densities on the bridges?

And so on and so forth. The list of questions that one can envision is endless—and not new. But with a laboratory like *TRANSIMS* at our disposal, for perhaps the first time ever, we do not have to build expensive bridges or do possibly dangerous tinkering with the system in order to answer them. Before leaving Albuquerqia, let's look at just one more example of *TRANSIMS*.

We began this section with the sad tale of the traffic crunch on Louisiana Boulevard brought on by Saturday-afternoon Christmas shoppers. A fairly obvious question about this dismal situation is, How many cars per hour have to enter the stretch connecting the two shopping malls before total gridlock sets in? By using *TRANSIMS* to introduce different levels of inflow and observing the result, we can answer this question

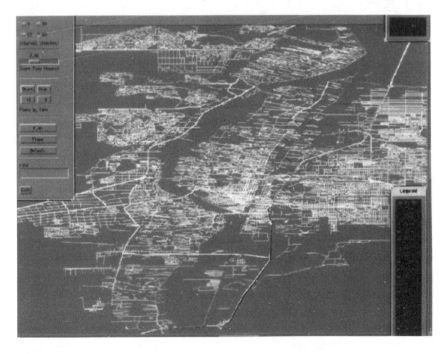

**Figure 4.6**   Traffic densities at 7:44 A.M. (see Plate VIII).

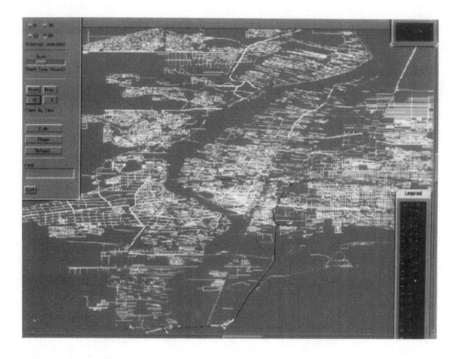

**Figure 4.7**    Traffic densities at 8:44 A.M. (see Plate IX).

quite easily. To illustrate, Figure 4.8 shows a space–time flow diagram when 3,000 cars per half hour enter this stretch of road. The figure should be interpreted in the following manner. A car enters at the upper left-hand corner of the diagram. Time increases as one moves down the vertical axis, while the road segment goes along the horizontal from left to right. Thus, the trajectory in space and time of a car entering the section and moving through it at a constant velocity appears as a diagonal straight line pointing downward from the upper left to the lower right of the diagram, the negative of the slope of this line being the reciprocal of the car's velocity. On the other hand, a car entering the system and coming to a standstill in a traffic jam contains a vertical straight line segment as part of its path in the diagram. This is because for such a car there is a period during which time marches on—but the car doesn't.

This picture allows us to see the appearance and disappearance of traffic congestion. Moving across the figure from left to right, there are what look to be regions of compression, in which the cars come together, followed by regions of expansion where the density of traffic is

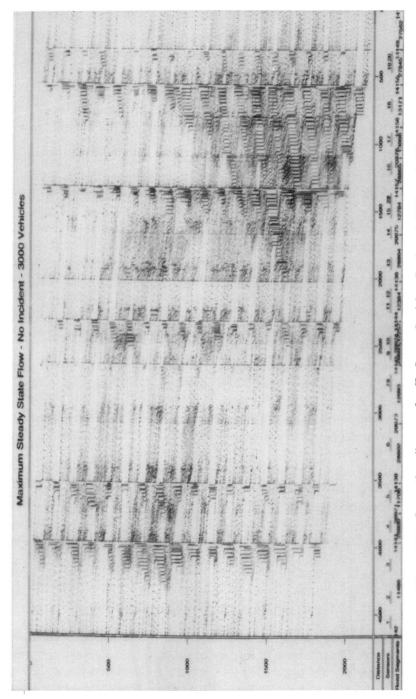

**Figure 4.8**  Space–time diagram of traffic flow on Louisiana Boulevard, 3000 cars/half hour.

relatively low. Some of these compression regions correspond to cross streets where there are traffic lights. But others, like the high-density region near the lower right-hand side of the figure, do not correspond to cross streets, traffic lights, or any other obvious source of congestion. Rather, they are what we have called emergent phenomena earlier in the book, that is, systemic properties arising out of the interaction of the vehicles cruising down this stretch of Louisiana Boulevard.

Here we are looking at a traffic jam in the making. To force the system beyond its capacity to cope, let's insert even more cars into the road segment and see what happens. Figure 4.9 shows the effect of raising the inflow of cars by 50 percent, from 3,000 cars per half hour to 4,500. Now the situation is completely supercritical. Almost as soon as the cars enter on the left we see traffic at a standstill, as evidenced by the patterns of vertical lines in the figure. Moreover, the stop-and-go nature that is characteristic of traffic jams is manifestly present, because the diagram shows almost no regions of rarefaction where cars can actually move. Paradoxically, the big congestion near the right side of the diagram that was seen when the inflow was 3,000 cars per half hour now smooths out somewhat when we increase the inflow to 4,500. Most likely, this is due to the inability of traffic to move on the road segment, thereby preventing enough cars from getting down the road to create the kind of jam seen earlier.

While one could write several books about the possibilities for using *TRANSIMS* to study road traffic networks, I think these examples convey the overall idea of what can be done with the system. Instead of belaboring this point, let us turn our attention to more general issues surrounding these types of artificial worlds, considering now not only their structural components, but also the actions of the agents making up such a world, how they choose these actions, and, most importantly, how the interactions of the agents give rise to emergent behavior in the system as a whole.

## Microworlds, Macrobehaviors

Speaking literally, the playing field on which the action takes place in all worlds like Albuquerquia is the memory bank inside a computing machine. Similarly, the agents in these worlds—traders in a stock market or drivers in a road-traffic system—are simply programs that interact

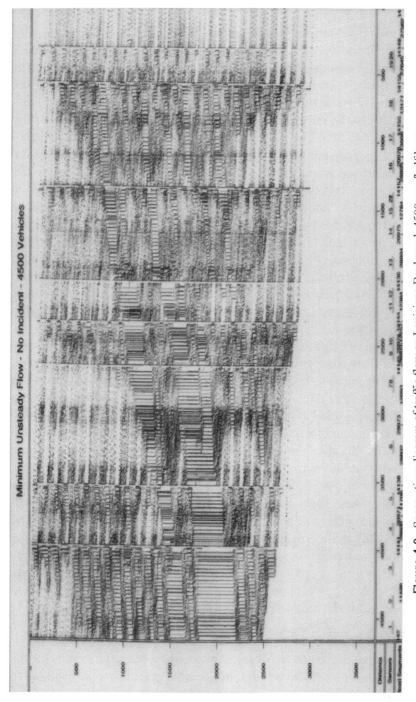

**Figure 4.9** Space–time diagram of traffic flow on Louisiana Boulevard, 4500 cars/half hour.

with each other by moving bits and bytes of data around from one memory location to another. If we take the environment and agents literally as being the memory bank of a machine and strings of computer code, respectively, we could end up in a world in which the agents compete against each other in a game dubbed *Corewar* some years back by *Scientific American* columnist Kee Dewdney.

According to Dewdney, Corewar was inspired by a story he heard about a programmer in a large company, who wrote a program called *CREEPER* that would create a copy of itself every time it was run. The program also had the ability to move from one computer to another in the company's computer network, which meant that it didn't take long before there were so many copies of *CREEPER* in the system that it began crowding out more useful programs and data. This creeping disease was finally stomped out when some bright light in the firm had the idea of fighting fire with fire. He wrote a second self-replicating program called *REAPER*, whose purpose was to seek out and destroy copies of *CREEPER*. When *REAPER* could find no more copies of *CREEPER* to destroy, it was designed to liquidate itself.

It was from this possibly apocryphal story that Dewdney got the idea for *Corewar,* a game in which two or more computer programs stalk each other in the memory banks of the machine, each program trying to destroy the other programs by taking over memory locations and hitting the other programs in vulnerable areas. A battle starts by dropping the programs into a block of the computer's memory at randomly selected locations. Initially, none of the programs knows where the others are located. The programs then take turns executing their instructions, one instruction at a time. Because the competing programs are the agents in the *Corewar* world, the "rules" these agents can use in their quest to destroy the ability of the other agents to survive are worth a closer look.

At any cycle, a program may execute whatever instruction its programmer has built in to it. The aim is to disable another program by ruining its instructions and thus forcing it to crash. Myriad strategies are possible. A program might take an offensive stance by standing back and lobbing software "bombs" into various memory locations, thus hoping to hit another program in a soft spot. On the other hand, a defensive strategy might be adopted in which a program repairs any damage it has suffered or moves out of the way when it comes under attack. The battle ends for a particular program when the operating system comes to an instruction in that program that cannot be executed. The program is then declared a "casualty of battle" and retires to the sidelines. The battle then continues with the remaining programs until only one of them is left unscathed.

The actual language in which *Corewar* programs are written is a variant of assembly language, which occupies a place somewhere between the purely 0/1 expressions of the language computing machines understand and the nearly natural-language statements of a high-level programming language like Basic, C++, or Fortran. Although it is not at all important here to understand the intricacies of the *Corewar* language, Table 4.1 gives a sample of the instructions that can be used to compose a *Corewar* program.

The environment in which the *Corewar* agents carry on their silent, deadly battles is called the Corewar Colosseum. It consists of 8,192 memory locations, of which the competing programs occupy just a small fraction—initially. Because the battle would be of considerably less interest if the programs knew where the other programs were located in memory, the method used for calculating a position is called *relative addressing*. So, for example, the instruction MOV 5 200 tells the simulator to go forward five address locations from its current position, copy what it finds in that location, go forward 200 addresses beyond the MOV instruction, and replace whatever it finds there with what it just copied. This example illustrates what is called direct mode, which means that the inputs 5 and 200 are interpreted as addresses to be acted upon directly. The *Corewar* simulator also admits what are called indirect mode and immediate mode for arguments. For example, in the indirect mode instruction MOV @5 200, the integer to be placed in relative address 200 is not the one found at relative address 5, but rather the one found at the address specified by the contents of relative address 5. Prefixing an argument with the # sign makes an argument immediate,

**Table 4.1**  Some instructions in *Corewar* programs.

| Instruction | Mnemonic | Arguments | Explanation |
|---|---|---|---|
| Move | MOV | A B | Move contents of address A to address B. |
| Add | ADD | A B | Add contents of address A to address B. |
| Jump | JMP | A | Transfer control to address A. |

**Table 4.2**  The first few cycles of *DWARF.*

| Address | Cycle 1 | | | Cycle 2 | | | Cycle 3 | | |
|---------|---------|-----|-----|---------|-----|-----|---------|-----|-----|
| 0 | | | | | | | | | |
| 1 | DAT | | −1 | DAT | | 4 | DAT | | 14 |
| 2 | ADD | #5 | −1 | ADD | #5 | −1 | ADD | #5 | −1 |
| 3 | MOV | #0 | @−2 | MOV | #0 | @−2 | MOV | #0 | @−2 |
| 4 | JMP | −2 | | JMP | −2 | | JMP | −2 | |
| 5 | | | | — | 0 | | — | 0 | |
| 6 | | | | | | | | | |
| 7 | | | | | | | | | |
| 8 | | | | | | | | | |
| 9 | | | | | | | | | |
| 10 | | | | | | | — | 0 | |
| 11 | | | | | | | | | |
| 12 | | | | | | | | | |
| 13 | | | | | | | | | |
| 14 | | | | | | | | | |
| 15 | | | | | | | — | 0 | |
| 16 | | | | | | | | | |

in the sense that it is then regarded not as an address location but simply as an integer. So the instruction MOV #5 200 results in placing the integer 5 in relative address 200.

To illustrate these instructions in action, Table 4.2 shows the first few cycles of the program called *DWARF,* which is a very primitive—but very dangerous—program that goes through the battlefield dropping a bomb on every fifth memory location by placing a 0 in it. Because 0 is the integer signifying a nonexecutable data statement, a 0 dropped into an enemy agent's program can bring it to a screeching halt.

Table 4.2 assumes that *DWARF* occupies memory addresses 1–4. At the outset, the program places the data −1 in location 1. Execution of the program begins with the next statement, ADD #5 −1. This has the effect of adding 5 to the contents of the previous address, which transforms the DAT −1 into DAT 4. Next, the program executes the instruction at absolute address 3, which is MOV #0 @−2. This moves the integer 0, specified as an immediate value, to the target address, calculated by counting back two addresses from address 3, arriving at address 1. The data value there, 4, is then interpreted as the address relative to the current position where the 0 is to be placed. In other words, the program counts four locations forward from address 1 and, hence, deposits a 0 at address 5.

The last instruction in *DWARF*, JMP −2, creates an infinite loop by directing execution back to absolute address 2. This again increments the DAT statement by 5, making its new absolute value DAT 9. Thus, in the next execution cycle a 0 is placed in absolute address 10. Subsequent 0s will then fall on addresses 15, 20, 25, and so forth.

From this description, it's easy to see that no program that stays put in memory and is longer than four instructions can avoid eventually being bombed by *DWARF*. There are only three ways for such a program to defend itself: (a) move to new locations, (b) stay put and repair whatever damage the bombs from *DWARF* wreak on the program, or (c) get *DWARF* first. The last option is a risky one, however, because the program has no idea where *DWARF* is located in memory. Unfortunately, there is no room here to go into the details of how to implement either of the first two strategies, so I'll have to refer the interested reader to the full account of the situation given in Dewdney's book cited in the references.

Before closing the door on *Corewar*, though, let's look at a battle between *DWARF* and the even simpler-minded program *IMP*, which consists of the single statement MOV 0 1. This means that it copies the contents of relative address 0 (that is, MOV 0 1) to the next address, relative address 1. As the program is executed, it moves through the array, one address per cycle, leaving behind a trail of MOV 0 1 statements. Thus, *IMP* is an example of a program that can move itself to new locations in the memory array. Now let's drop *DWARF* and *IMP* onto the floor of the Corewar Colosseum and see what happens.

Initially, it might seem that the volley of 0s lobbed into every fifth location by *DWARF* would move through the memory array faster than *IMP* moves. In fact, this is the case, but it does not imply that *DWARF* necessarily has the advantage. What is at issue is whether or not *DWARF* will fatally strike *IMP*, even if the barrage of 0s does catch up. That this might not be the case is evident, since the very nature of *IMP* causes it to move copies of itself to new locations. Thus, even if *DWARF* does ruin some of these copies it will not be victorious in the battle unless it manages to wreck *all* of them.

On the other hand, if *IMP* reaches *DWARF* first, it is likely that *DWARF* will be subverted and become a second *IMP*. This is because

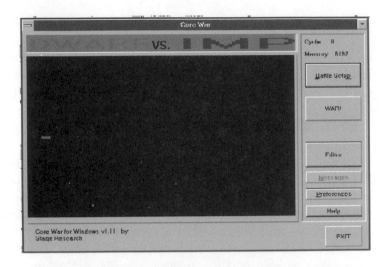

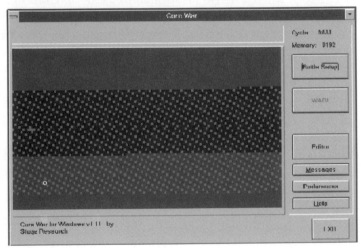

**Figure 4.10** *DWARF* versus *IMP* battle: the beginning (top) and the end (bottom).

when *DWARF*'s JMP −2 instruction transfers execution back two locations, the instruction found there will be the single MOV 0 1 instruction constituting *IMP,* which in effect transforms *DWARF* into *IMP.* In this event, the battle turns into two copies of *IMP,* chasing each other around the colosseum. And under the rules of *Corewar,* the battle ends up in a draw, but this may not necessarily happen, as seen in Figure 4.10. Here the top of the figure shows the battlefield at the beginning, and the bottom panel shows what things look like at the end: a victory by *DWARF.*

The foregoing discussion touches only the tip of the iceberg of the depth and sophistication of *Corewar* as a paradise for hacker's testing strategies under battlefield conditions. Space precludes further discussion here. Extensive treatment can be found in the articles and software pointers given in the references. For now, let us reflect on what *Corewar* has to tell us about the structure and use of would-be worlds.

The most important point is to note that we have been speaking of both the environment and the agents in anthropomorphic terms, using words like *battlefield, Colosseum, players, warriors,* and the like. In reality there is nothing here but memory locations in a physical device, the computer, together with information structures creating and destroying what amount to ON/OFF patterns in this memory array. So, for instance, the patterns shown in Figure 4.10 are simply a map of the memory array, showing which address locations are occupied at that moment by the two warriors, *DWARF* and *IMP.* This type of memory map differs in no essential way from the ON/OFF patterns of an array of light bulbs in, say, a Times Square message board. This means that in order to see such patterns as having significance in terms of, for instance, traffic congestion in Albuquerquia, they must be *interpreted.* In their native state, the patterns have no meaning at all; they are pure syntax, just long strings of 0s and 1s in the computer's memory. It is the act of interpretation, the injection of semantics, so to speak, that allows these electronic worlds to make contact with their real-world counterpart(s).

The second key issue is suggested by the nature of the two *Corewar* warriors, the programs *DWARF* and *IMP:* Both are very stupid—but extremely dangerous. They are stupid because the programs take no cognizance of anything that might be learned about the location or nature of their opponent. *IMP* simply moves ahead blindly, one location at a time, whereas *DWARF* just sits back and blithely lobs bombs into every fifth memory location. Thus, like more than a few of the managers of companies and research institutes I've known, both programs operate on the basis of myopic, totally rigid and insensitive protocols, never learning anything about their environment or the agents they are dealing with. Stupid—but dangerous.

We are all born stupid. But the commonly accepted method for climbing out of the pit of stupidity is for individuals to learn and for populations to evolve. So let's spend a few pages discussing how agents in would-be worlds accomplish these most human of tasks.

## How Life Learns to Live

One of the few things upon which the scientific community seems to be in total agreement is that learning takes place in the brain. Exactly *how* it takes place is, however, quite another matter, one for which there is nothing remotely approaching a consensus. To understand the difficulties in creating a viable theory of learning, let us consider briefly the structure of the brain and how it seems to function.

The brain is built from an unimaginably large collection of *neurons,* which communicate with each other by means of a dense web of connections. The "wires" in this connective network are called *axons* and *dendrites,* and the points of connection between one neuron and another are termed *synapses.* The dendrites are the input channels to the neuron, along which electrochemical pulses flow from other neurons in the network. Some of these pulses constitute *excitatory* inputs, while others are *inhibitory.* If the difference between the number of excitatory and inhibitory inputs exceeds a certain threshold, the neuron then fires a pulse that travels down the axon. The pulse serves, in turn, as either an excitatory or an inhibitory input to other neurons. Figure 4.11 shows the structure of a typical neuron.

The diagram in Figure 4.12 (page 152), which is taken from a slice of real brain tissue, gives some indication of the complexity of the network of neurons forming the human brain. Here we see three small neurons (A, B, and C), two medium-sized pyramidal neurons (D and E) and three other types of neuronal cells (F, J, and K), together with each cell's respective synaptic connections. Note that the objects marked G, H, and I are dendrites from neurons lying at a level deeper than the slice depicted in the figure. Because the cortex of the human brain consists of around 100 billion such cells, we see that this diagram is indeed an infinitesimally thin slice of the entire brain.

The fact that the brain as a whole consists of about 20 billion copies of Figure 4.12 is what accounts for our ability to learn, speak, and, in general, think. Consider how this network of neurons, which after all is basically just a set of interconnected ON/OFF switches, can accomplish such miraculous feats.

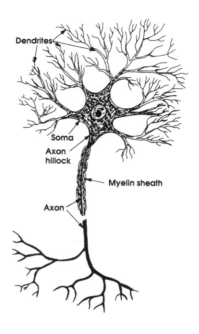

**Figure 4.11** A typical neuron.

In 1949, Donald Hebb published the volume *The Organization of Behavior,* in which he proposed the theory that if pairs of neurons are active (i.e., firing) at the same time, the connections between those two neurons is strengthened, thereby reinforcing those particular pathways in the brain. According to Hebb's theory, two neurons are considered to be active simultaneously when the sending neuron's pulse occurs at very nearly the same time as the receiving neuron's output pulse. So, according to Hebb the processes of learning and memory storage involve changes in the strength with which neuronal signals are transmitted across individual synapses.

In the 40 years or so since Hebb's work, evidence has piled up in favor of this theory of learning at the neuronal level. It is indeed at the synapses where neural tissue is most significantly changed as a result of signal transmission. The strength of the connection between neurons can be dramatically altered by the chemical nature, size, and/or form of the synapse. Current thinking about memory suggests that it is changes in the number of active chemical receptors at the receiving site that is the main source of memory in adults. Some of these changes may last for only a few seconds or minutes, whereas others can be permanent.

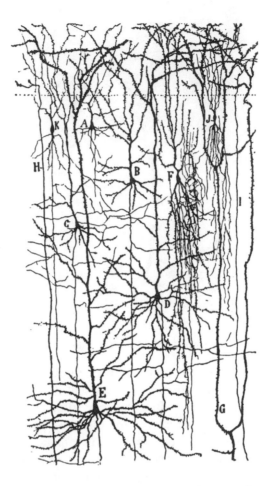

**Figure 4.12** Neurons and their synaptic connections.

Unfortunately, it is extremely difficult to gather experimental evidence to validate theories of specific synaptic changes. The problem is that the changes are spread out over many neurons, and at each synapse the change is rather small. Moreover, the chemical receptor molecules are very quickly redistributed among the synapses of the same cell according to demand. Nevertheless, the Hebb theory has stood up rather well as a starting point for how the machinery of the brain works to give rise to behavioral phenomena like memory and learning.

The overall structure of the brain is composed of many interesting types of neural cells, which are themselves arranged into a variety of

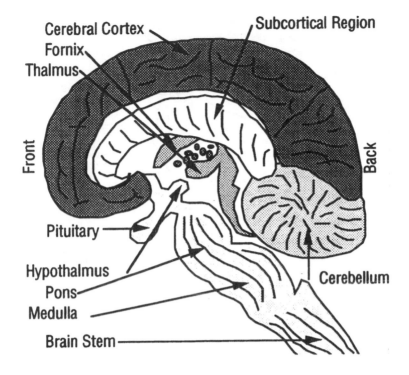

**Figure 4.13**    The human brain.

regions. These regions are depicted in Figure 4.13. Let's focus our attention only on the cerebral cortex, which is the part of the brain that is of greatest importance for cognitive activity.

The cortex is a continuously folding layer forming the outside of the brain. In humans, this region is often termed the *neocortex,* and it is the newest part of the brain, evolutionarily speaking. It is the part of the brain where reasoning and thought occur. Although the cortex can be divided into a great many areas, both structurally and functionally, as indicated in Figure 4.14, all the parts are built of the same basic components and are linked together in similar ways. The functional differences among areas are probably due to the different sensory signals coming into them, not to a difference of structure.

In the highest levels of the central nervous system, the neurons form many thin layers called the *gray matter.* In this region the firing threshold of the neurons tends to be higher than for neurons in the rest of the brain, probably as a way of suppressing spurious "noise" from other sources,

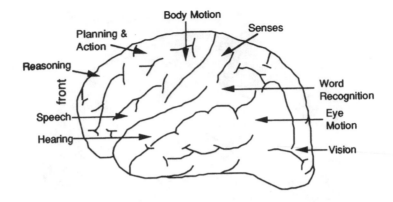

**Figure 4.14** A map of the functional areas of the brain.

thereby preventing sporadic, unneeded (and unwanted) responses. This, in turn, acts to promote the overall stability of the entire cortical system.

The human cortex is a pleated sheet of around 2,000 square centimeters (approximately 2 square feet), which is 2 to 3 millimeters thick. This sheet contains about 100 billion cells and several times as many synapses. Most of the human cortex is made of many layers of densely interconnected neurons, each neuron connected to between 1,000 and 100,000 other neurons. Evidence suggests that the smallest functional unit in the brain is not actually the single neuron but rather a collection of 4,000 or so neurons forming a column. This is a cylinder 300 microns across, which vertically connects six layers of the cortex. These cylinders, or *modules,* are connected to each other so that the brain can act as a single, integrated system.

The essence of arguments in support of "machines who think" is that a computing machine, for example, a Turing machine of the sort described earlier, can somehow be made to perform the same functions as the brain. This argument is immeasurably strengthened if we can mimic in a machine, not only the functions of the brain, but also its actual logical structure. Let us examine how the computer-science community has gone about trying to accomplish this trick.

## Neural Nets

In 1943, Warren McCulloch, a neurophysiologist at the University of Illinois, and Walter Pitts, a graduate student in mathematics at the

University of Chicago, published a path-breaking article titled, "A Logical Calculus of Ideas Immanent in Nervous Activity." Their basic result was astonishingly simple: the operations of a neuron and its connections with other neurons, what we now call a *neural network*, could be modeled purely in terms of the operations of mathematical logic. This model considers a neuron as being activated and then firing another neuron in the same way that a proposition in a logical sequence can imply some other proposition. Furthermore, the analogy between neurons and logic can be pictured in electrical terms as signals that either pass—or fail to pass—through a circuit. From here it is but a small step to implementing the logical structure created by McCulloch and Pitts in the hardware of a digital computing machine. In point of fact, what McCulloch and Pitts provided was nothing short of what brain scientist Michael Arbib has termed the "physiology of the computable," demonstrating that each Turing machine program can be implemented using a finite network of McCulloch–Pitts logical neurons. Let's briefly review the general setup created by McCulloch and Pitts.

A McCulloch–Pitts neuron is an object with, say, $m$ input channels ("synapses") and one output channel (the "axon"). It is thus characterized by $m + 1$ numbers: the threshold level at which the neuron "fires," and the weights associated with the $m$ input lines. These weights are used to determine the total impulse arriving at the neuron from its various input channels by just adding up the pulses on each input channel, biasing each pulse by the weight assigned to its input line. The neuron will then fire a pulse along its single output channel (the axon) at time $t + 1$ if this weighted sum exceeds the threshold level. Figure 4.15 shows such a neuron schematically, where the weights on the input lines are labeled $w_i$ and the threshold level is denoted by $\theta$.

A *neural net* consists of layers of McCulloch–Pitts neurons that are connected to each other. In such a network, a neuron can be located in one of three layers: the input layer, the output layer, or one of the hidden

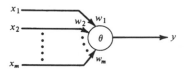

**Figure 4.15**   A McCulloch–Pitts neuron.

layers. The input neurons receive information from the outside world and send it to neurons in the hidden layers. The output neurons then give the network's response to the inputs by either firing or remaining silent. This pattern of firings is then interpreted as things like letters, pictures, or concepts. The schematic diagram in Figure 4.16 shows the layout of such a network, where each large circle represents one McCulloch–Pitts neuron, and the various lines are the synaptic connections linking them.

To illustrate the operation of such a network, suppose we want to train the network to recognize characters of the alphabet. First of all, we code the possible firing patterns at the output terminals to correspond to the various letters of the alphabet. For instance, the pattern consisting of the first neuron firing and all others not firing might correspond to the letter A, while the first and second neurons firing and all others not firing could represent B and so on. We then write a letter on, say, a digitizing tablet, which sends signals to certain input neurons. These neurons, in turn, send signals to the hidden layers. These hidden neurons then transmit a signal to the output layer of neurons, which will cause a certain output pattern to appear. If this pattern corresponds to the letter shown at the input, fine; otherwise, we employ some procedure (there are many possibilities here) to adjust the weights on the various synapses and repeat the process. If a pattern is eventually formed at the output terminals that agrees with the letter that was shown at the input, we then say that the network has been "trained" to recognize that letter.

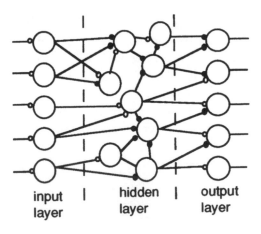

input layer | hidden layer | output layer

**Figure 4.16** A neural network.

It is important to note here that when the network has been trained to recognize, say, the pattern A, this knowledge is distributed throughout the network. Basically, the network's knowledge of the letter A resides in the weights on the various input lines (synapses). If we then show a slightly distorted version of the original A at the input, the network will still be able to recognize it as having a family resemblance to the platonic letter A. Even more importantly, the network's performance degrades only gradually as the weights are modified just a bit, either by noise from the outside world or, perhaps, by complete excision of neurons from the network (which effectively sets one or more of the weights to zero).

This description of a neural network is the bare-bones minimum, and there are a lot of ways to jazz it up to make it conform a bit more closely to the actual material structure of the brain. In fact, the description just given has abstracted away just about *everything* associated with the behavior of real neurons except for their all-or-none firing behavior. Even with this simplest of all possible neural networks, it is already possible to prove that any computer can be simulated by a network of such neurons. Thus, a neural network has exactly the same computational capability as a Turing machine. The importance of this surprising fact is that it shows we can carry out highly detailed and complex computations with very simple components. The critical factor is not the complexity of the components, but the way the components are linked together. In short, it's connections that count, not components.

The ability of a neural net to both compute anything that can be computed and to learn to recognize patterns is what gave rise to Alan Turing's hope in the early 1950s for the creation of a genuine "thinking machine," one that would think just like you and me. The jury is still out on this gleam in Turing's eye, as you will discover by consulting the material cited in the references for this chapter.

Having now seen how an individual brain might operate and learn new things about the environment in which it lives, we now turn our attention to how computer scientists mimic the idea of learning, not in individuals, but rather by evolutionary change in *populations* of individuals. This is important because evolutionary changes can only occur in a population of organisms. If we want to study learning as an evolutionary process, we have to abandon the individual and direct our attention to how changes take place in a group.

## The Elements of Evolution

Fortune's formula for survival dictates that those members of a population that most well adapt to their environment will be the winners in the evolutionary sweepstakes. Put more compactly, we can state this principle as

$$adaptation = mutation + variation + selection.$$

Here is a bit more detail on these three components that make up the neo-Darwinian view of evolutionary processes.

- *Mutation:* The characteristics determining an organism's fitness are determined by the genetic makeup bequeathed to the individual by its parent(s). That genetic endowment can be modified before transmission to the individual's offspring by chance events (point mutations), as well as by genetic crossover (sexual combination of different genes).
- *Variation:* The members of the population are capable of manufacturing good—but not perfect—copies of themselves. In particular, parents pass on copies of their mutated genes to their offspring.
- *Selection:* Members of the population having good genetic makeups are able to produce more offspring, on the average, than those possessing bad genes.

From these simple assumptions flow the bewildering variety of animal and plant species inhabiting the earth today. Computer scientists have abstracted these very same components of the evolutionary process to create what are termed *genetic algorithms* for solving difficult problems, as for example, finding the shortest path through a network. Let's see how this works.

## Genetic Algorithms

Suppose you have a robot that must traverse the maze shown in Figure 4.17. At any given step, the robot can move **Backward**, **Forward**, **Right**, or **Left**, actions that we can code as pairs of binary digits in the following fashion: **B** = 00, **F** = 10, **R** = 11, **L** = 01. Assume the task is to find a sequence of moves that will take the robot from the entrance to the exit in no more than, say, 20 steps.

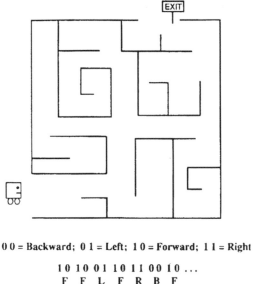

0 0 = Backward;  0 1 = Left;  1 0 = Forward;  1 1 = Right

1 0  1 0  0 1  1 0  1 1  0 0  1 0 ...
F    F    L    F    R    B    F

**Figure 4.17**   Navigation problem for a robot.

Because each sequence of 20 steps consists of a string of length 20 composed of the symbols **B, F, R,** and **L,** each of which requires two binary digits to represent it, we can express each such sequence of 20 steps as a binary string 40 bits long. A partial string of this type is shown in Figure 4.17. We can think of these strings as being the "genotypes" of different "organisms" in a population, each of whose members is a particular sequence of moves in the maze. These organisms compete with each other to see which of them represent the best way to get the robot from the entrance of the maze to the exit. As a measure of the fitness of any such individual, we simply let the robot follow that sequence for 20 steps, and then calculate the minimal number of steps from its final position to the exit. The smaller this number, the greater the fitness of the sequence. Maximum fitness 0 is achieved with any sequence that causes the robot to reach the exit within the allowed 20 steps. How do we find such a maximally fit sequence? We do it the same way nature does, by finding a procedure for searching the space of possible genotypes for those of high fitness. Such a procedure is called a *genetic algorithm* (GA).

The key elements in a GA are abstractions of the processes of mutation, genetic crossover, and selection. Basically, we have to find a way to create new genotypes from old, and then be able to evaluate the relative fitness of any genotype in the population. As noted earlier, there are two ways of modifying existing genotypes to get new ones. They are

1. *Mutation:* Because each genotype in the robot problem is represented by a string of 40 bits, we can randomly flip some bits in any string to create a new string. For instance, the string segment 001010101011 might be mutated in its third position to get the offspring 000010101011. Mutation can, in principle, occur at every position in a string with some probability (usually low, say, 1 in a 1,000).

2. *Crossover:* This operation interchanges subsequences of two genotypes to create two offspring. For example, suppose the two "parent" genotypes consist of the following subsequences of length 12:

$$A = 010011010100 \quad \text{and} \quad B = 010111100010$$

Now let each of these subsequences be cut after position 6. The second half of $B$ could then be used as the first half of a new genotype $A'$, whose second half is the old second half of $A$. This yields $A' = 100010010011$. Similarly, the second half of $A$ could be used to form the first half of the offspring $B'$, along with the first half of $B$ to get $B' = 010100010111$. This type of crossover operation mimics (roughly!) the biological operation of genetic recombination between two organisms, those whose genetic strings come only in single strands. (Technically, these are what are termed haploid organisms. We will not discuss diploid organisms here, whose genetic makeup is expressed in pairs.)

Mutation and crossover are the two main ways GAs mimic the biological operations of genetic variation. Now what about selection? This is probably the easiest part of the adaptation equation—for computer scientists, anyway. The selection operation simply chooses those genes in the population that will leave copies of themselves in the next generation. The fitter the gene, the more likely it is to be chosen to reproduce. That's all there is to it.

To summarize, a typical GA proceeds as follows. Given a particular problem to be solved, we first devise some scheme whereby candidate solutions can be coded as strings of 0s and 1s. We then

1. Randomly generate a population of, say, $N$ genes, each gene representing a possible solution to the problem.

2. Calculate the fitness of each gene in the population.

3. Go through the following steps (a)–(c) until $N$ new genes (offspring) have been created:

   a. Select a pair of genes (parents) from the current population, such that the likelihood of any gene being selected increases with higher fitness. Selection is done here with replacement, meaning that if a gene is selected, it can be selected again at this same step.

   b. With a given probability $p$, crossover the chosen pair, the crossover location being selected with equal likelihood for all locations. With probability $1 - p$ there is no crossover, in which case two offspring are formed that are exact copies of their parents.

   c. Mutate the two offspring ("flip a bit" from 0 to 1 or vice versa) with a fixed likelihood at each location, and place the resulting gene in the new population.

4. Replace the current population with the new population.

5. Go to step 2.

Each repetition of these steps constitutes one generation of the algorithm. A typical run of the GA consists of 100–500 generations, after which there are often one or more very fit genes in the population. The most fit of these candidate solutions is then taken to constitute a solution to the original problem. So, for instance, using this procedure for the robot's navigation problem shown in Figure 4.17, it doesn't take long to arrive at a gene of highest possible fitness 0, meaning in this case that the path actually ends up at the exit to the maze. Decoding this optimally fit string yields a route like **FLRFL** that takes the robot through the maze.

Now we know how to make agents get smart as individuals (by neural networks) and how populations get smart (by following the principle of natural selection). Now we will look at how these ideas work in a would-be world whose agents are themselves surrogates for *real* biological genes.

## *Tierra*

January 4, 1990, a day to be remembered: the day when the first noncarbon-based life form came bubbling up out of the computing machine of Tom Ray, a naturalist from the University of Delaware. Ray, a soft-spoken, unassuming man, had for years been puzzled by what it is *exactly* that makes life on earth special. Unfortunately, earthly biology is based on a sample size of 1, and as yet we have no examples of extraterrestrial life with which to compare it. Until 1990 there seemed to be no way to separate life as we know it here on earth into its contingent and necessary parts, to discover what is general and what is peculiar about life.

Because interplanetary travel seems impractical, at least for the time being, Ray took the bold step of considering how he might create an alternate life form in his computer. In contrast to many modeling exercises of a similar nature that had been carried out in the past, Ray's goal was not to shed light directly on natural life. Instead his aim was to create life forms radically different from those we see around us, forms based on a completely different physics and chemistry. He planned to then let these forms evolve in their silicon environment, generating their own characteristic phylogeny. These independent living organisms would then serve as a basis for a comparative biology, hopefully shedding light on what is special and what is general about earthly biology.

In putting his experiment together, Ray was keenly aware of the fundamental difference between simulating life and synthesizing it. His plan was to start with an "ancestral" organism tailor-made to be capable of self-replication and open-ended evolution. Here *open-ended* means that the structures or processes that might evolve from the ancestral organism are not taken from a set of possibilities predefined by the experimenter/creator. Rather, they are somehow able to emerge solely through the workings of the evolutionary process itself. Prior to Ray's work, evolutionary programs and models were not open-ended in this sense, thereby severely limiting their potential for development of interesting, new structures because in such experiments the organism's genome is taken from a set of *predefined* possibilities. Moreover, the entities in these earlier exercises were assumed to evolve in accordance with criteria of mutation, crossover, selection, and replication designed into the simulator by the modeler. These sorts of *simulations* of life are closed-ended and dead. What Ray wanted was a *synthesis* of life that would be open-ended and alive, so the overriding consideration was to avoid falling into this kind of representational cul-de-sac.

We will turn in a moment to a discussion of how Ray managed to create his ancestral organism capable of replication and open-ended evolution. Once this ancestral organism was available, the next step of the experiment was to drop it into an environmental "soup," where it would find the resources needed to start the reproductive and evolutionary process in motion. With some luck, Ray hoped that after several generations he would find a rich diversity of different electronic organisms frolicking about in the playing fields of his computer's memory, a profusion of life-forms that would mirror the explosive appearance of multitudes of species that arose here on earth during the Cambrian era around 600 million years ago. The focus of Ray's experimental world (which he named *Tierra,* the Spanish word for Earth) was not to study the processes leading to the origin of life, but rather to look at how the vast diversity of species of organisms exemplified by the Cambrian explosion could have come about through the simple process of adaptation. Now we'll take a look at how Ray managed to pull off this modeling tour de force.

### Digital Organisms and Silicon Environments

In the everyday world, life makes use of energy from the sun to organize matter. These forms of organized matter then carry out their activities in the physical environment of the earth. By way of analogy, the organisms in *Tierra,* which take the form of self-reproducing computer programs, make use of time on the computer's central processing unit (CPU) to organize the machine's memory locations. In the natural world, life evolves as organisms compete for food, shelter, mating partners and all the other things that natural selection acts upon. Those genotypes leaving the most descendants increase in frequency as time goes on, while less-fit members of the population leave fewer offspring and their genotypes eventually dwindle in number until the entire species goes extinct. In the world of *Tierra,* digital life goes through the same process, as self-reproducing programs compete for CPU time and memory. These programs/organisms evolve strategies to exploit one another, with the programs that get more time and memory allocation being able to leave more copies of themselves in the next generation. Now let us be a little more specific about what these *Tierra* creatures really are, and how this evolution in silicon actually works.

In *Tierra,* the computer's CPU and memory constitute the physical environment, or playing field, upon which the evolutionary process unfolds. A Tierran organism consists of a self-replicating program written in basic language of the machine. (Technical aside: These programs are actually written in assembler language, which is one step removed from the basic machine code of 0s and 1s that all computers eventually operate with. Biologically speaking, these lowest-level machine instructions are analogous to the amino acids making up each and every protein of all living things.) Thus, the Tierrans are digital versions of the creatures of the RNA world, because in *Tierra* the same structures carry both the genetic information and execute the metabolic activity of the organism. This is in contrast to the DNA-based world of modern organisms, in which these two functional activities are carried out by different structures (the DNA/RNA and the proteins, respectively).

In order to prevent these digital organisms from gaining access to the actual hardware of the machine they inhabit, Ray arranged for the entire *Tierra* system to run on what is called a "virtual computer" inside the physical machine. What this means is that inside the real computer he emulated *in software* a computer that hosts the *Tierra* world. In other words, Ray created a set of *software* instructions that mimic the operation of a physical *hardware* machine. As far as the real-world computing machine is concerned, the Tierran organisms are simply data that look no different than the data from a word processor, graphics package, or spreadsheet program. There are many reasons for doing this, perhaps the most important is that it prevents the self-replicating Tierran creatures from spreading from one machine to another and infecting a network with copies of themselves, much as a computer virus might do. Keeping them inside a virtual machine prevents these strange creatures from ever crawling up from inside the real machine and making mischief in the natural world. Other reasons favoring a virtual world for *Tierra* include the desirability of making the Tierran world independent of the hardware of any particular machine, and the fact that machine languages interpretable by real hardware are notoriously "brittle" (that is, almost every change made to such a program leads to a new program that causes the machine to crash). Using a virtual computer, whose virtual machine code is designed with evolution in mind, gets around this difficulty.

The operating system of the Tierran virtual computer determines communication among the organisms, allocation of CPU time, disposition of memory space, and all the other things constituting the operating environment for the digital organisms. The Tierran entities then evolve so as to exploit the features of the operating system to their advantage.

Here are the principal activities of the Tierran operating system:

**Memory Allocation (Cellularity).** The block of memory in the real computer occupied by the Tierran virtual computer is called the "soup." Typically, it consists of around 60,000 bytes of memory. Each inhabitant of *Tierra* constitutes some block of memory in this soup.

In order to maintain the integrity of organisms in the real world, evolution discovered the advantages of a semipermeable membrane enclosing the basic unit of life, which is the cell. An analog of the cell membrane for digital organisms is needed to prevent the activity of one organism from easily disrupting that of another. The Tierrans are protected in this sense by allowing only the organism itself to have the privilege of writing onto its block of memory. This means that other Tierran creatures can look at any other organism's structure—and even execute that organism's code—but only the organism itself can modify its own structure.

When a Tierran divides in the process of replication, the mother cell loses write privileges on the space of the child's cell, but the mother can then allocate another block of memory for further replication. At the moment of division, the offspring cell is given its own block of memory, and is free to allocate its own second block for its own replication.

**Time Sharing (The Slicer).** In order for each member of the Tierran population to go about its business at the same time as other members, the Tierran operating system should ideally function in parallel, meaning that the operations on each Tierran should take place at the same moment of time. This is not strictly possible with a serial-processing machine—real or virtual. However, the operating system does a good job of approximating a true multitasking environment by doling out small slices of CPU time to each Tierran creature in turn. As long as the slice size is small relative to the time of a generation of Tierran organisms, this time-sharing mode gives a good approximation to true parallelism.

The number of instructions of an organism's code that is executed on its slice of CPU time is set proportional to the size (total number of statements) of the organism's code. Bigger organisms (those with longer codes) get more of their instructions executed than smaller ones, subject to a weighting factor that is greater than, equal to, or less than 1. If this factor equals 1, the slicer is indifferent to size, so that the likelihood of an instruction being executed does not depend on the creature's size. On the other hand, setting the factor less than (greater than) 1 gives small (large) creatures more CPU cycles per instruction. This parameter determines whether selection favors small or large creatures, or is neutral with regard to size.

**Mortality (The Reaper).** It is clear that a population of self-replicating objects will eventually fill up a finite world unless there is some kind of death mechanism that removes organisms from the soup. In *Tierra* this mechanism is the reaper, which begins killing off creatures from a queue when the soup gets filled to some specified level, usually around 80 percent. The death of a Tierran involves taking back its memory allocation and removing it from both the Slicer and the Reaper queues. Note, however, that the dead creature's code is *not* removed from the soup. This means that other individuals in *Tierra* can make use of this dead code later on for their own purposes. Biologically, this leads to a kind of gene bank that organisms can draw upon to gain an evolutionary leg up on their cohorts.

The way the Reaper works is to put a creature at the bottom of a queue when it is born. Then at each cycle the Reaper kills the creature at the top of that queue. Tierran organisms can move either up or down in the queue according to how successful they are in executing certain instructions. In particular, when a creature executes an instruction that leads to an error condition, it moves up the queue one position as long as the individual above it in the queue has not accumulated even more errors. On the other hand, successful execution by a creature of difficult instructions that usually leads to an error moves the organism down the queue, subject to the same proviso about the number of errors accumulated thus far by the creature just below it in the queue. Thus, the net effect of the Reaper is to cause algorithms that are basically flawed to rise to the top of the queue and be killed off. Good, healthy algorithms, on the other hand, have a greater longevity. Eventually, though, the likelihood of death increases for all creatures with age.

*Variability* **(The Mutator).** Evolution cannot get started unless there is some way for the genome of a creature to be modified and some way for this modification to be passed along to the creature's offspring. In *Tierra* this happens in two different ways. The first corresponds to random mutations in the organism itself. Bits in the binary string representing the organism's program are randomly flipped from 0 to 1 or 1 to 0 at some fixed rate (for example, 1 bit flip for every 10,000 instructions executed). This is analogous to the mutations in real-world DNA caused by cosmic rays. The effect of bit-flipping is to prevent any creature from being immortal, because eventually every creature will have some of its code changed by this process and, hence, will eventually mutate to a form that is no longer viable.

Besides these kinds of random mutations to the organisms, the process of replication is also carried out imperfectly at some low rate. In this way, the behavior of a Tierran creature is not fully deterministic, but can vary unpredictably due to these low-frequency random mutations.

Now all the pieces of the *Tierra* world are in place. We have a spatial environment (memory) and a source of energy (CPU time), a method for allocating these resources to organisms (the Slicer), a way of keeping the population finite in a finite world (the Reaper), and a mechanism for evolution (the Mutator). All that is missing is an ancestral organism, an "Urtyp," to drop into the soup and get the evolutionary process off and running.

To create the Tierran Ancestor, Tom Ray wrote the only assembler-language program he has ever written in his life. It is nothing fancy; simply a self-replicating program that happens to have turned out to be 80 instructions in length. Ray originally wrote the Ancestor for use in debugging his experiment, and thus built no functional features into it other than the ability to self-replicate. On the magical night of January 3, 1990, Ray dropped the Ancestor into the soup and *Tierra* was born. What happened?

## Evolution with a Capital "E"

When Ray awoke the next morning and looked at *Tierra,* to his great surprise he found his electronic world literally teeming with creatures displaying a dazzling array of structures and activities. It was as if the

Cambrian explosion had taken place in a period of just a few hours rather than requiring the 3 billion years or so of evolutionary buildup before it took place here on earth. In summary, what Ray discovered swimming around inside the Tierran virtual computer after execution of 526 million instructions were creatures of 366 different size classes, 93 of which achieved subpopulations of five or more individuals. Because creatures of different sizes would have some difficulty engaging in genetic recombination if there were sexual reproduction, we can biologically interpret these size classes as different species. But this multitude of species was not all that evolution washed up on the shores of *Tierra*.

Besides simple mutations that do not differ in size from the Ancestor, the soup also contained objects that can only be described as parasites. These are creatures that, because of mutations in the replication procedure of their parent, do not contain the instructions needed to copy themselves. Thus, such creatures need to read the copying instructions from some host creature in the soup in order to replicate, and then use those instructions for their own reproduction. Such a freeloader (or parasite) will then be able to copy itself more frequently than objects containing the copying instructions, because the parasites are smaller in size, hence requiring less resources to generate offspring.

An evolutionary arms race between such a population of hosts and parasites is shown in Figure 4.18, in which each image represents a map of the computer memory array, a soup of 60,000 bytes divided into 60 segments of 1,000 bytes each. In the upper left, the hosts (red) are very common, while parasites (yellow) have just begun to appear. The upper right panel shows that hosts are now rare because parasites have become very common. Hosts immune to the parasites (blue) have now started to appear, but they are rather rare in the population. In the lower left, we see that immune hosts are increasing in frequency, relegating the parasites to the top part of memory. Finally, in the lower right panel we see that the immune hosts have taken over the population, whereas parasites and susceptible hosts have declined in frequency. The parasites will now soon be driven to extinction.

As in the real world of "wet-life" evolution, Tierran organisms eventually found ways to generate creatures immune to these parasites. In the space available here it is a little difficult to explain the exact mechanism for how this immunity comes about, but it is related to the size of the parasite and the size of the creatures whose copying instructions the

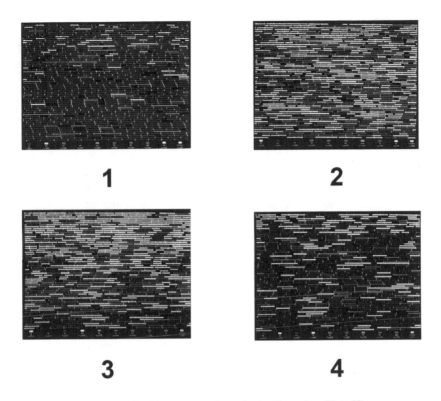

**1**　　　　　　**2**

**3**　　　　　　**4**

**Figure 4.18**　Emergence of species in *Tierra* (see Plate X).

parasite is trying to use. If these two sizes are incommensurate, all of a sudden the parasites begin having a rough time staying in the game. It turns out that even *hyperparasites* arise naturally, driving the parasites themselves to extinction. After awhile, the *Tierra* world even begins to evolve social organizations, which consist of communities with relatively high levels of genetic uniformity, implying high levels of genetic relatedness between individuals of the community.

Well, I think you get the idea. Just about all the features one sees in natural evolution and just about all the kinds of functional activities of organisms familiar from earthly life make their appearance in *Tierra*, as well. Moreover, the well-chronicled increase in complexity claimed for biological systems is also noted in the highly sophisticated activities of the evolved organisms. In *Tierra*, creatures that do not interact initially discover ways to exploit one another, as well as ways to avoid being exploited. Just as with the traders in Brian Arthur's stock market

169

discussed in chapter 2, the creatures in *Tierra* come to learn that they must continue to evolve to meet the challenges posed by other players in the game; hence, there is no one best way to survive in their constantly changing environment. Change is the name of the game.

It's of interest to note that in 1996, Andrew Pargellis of Bell Labs created an ecology of living forms *without* having to first seed the soup with an ancestral organism. Rather, Pargellis's experiments randomly combined segments of computer code until finally a self-replicating string appeared. From there on, the normal evolutionary mechanism took over and generated myriad different life-forms. Details on this world, which Pargellis calls *Amoeba,* are given in the material cited in the references.

In closing this discussion of *Tierra,* it is worth mentioning that all of Tom Ray's experiments took place within the confines of a single real computer. Because the CPU and memory of this machine sets the boundaries of the Tierran world, it's evident that the kinds of creatures that evolution might turn up are in some way limited by these hardware constraints. Recently, Ray has proposed setting up *Tierra* on the Internet in such a way that unused computer capacity in your machine and mine could be employed as part of the overall Tierran resource. Once he can enlist enough volunteers to donate computer resources to his digital "wildlife reserve," Ray will release his program to go forth and multiply. Under these conditions of a Tierran world, with vastly more time and energy resources, he expects to see the Ancestor's progeny displaying many more strategies for survival and reproduction than in the relatively simple environment provided by any single machine.

As noted, Tierran organisms evolve something akin to social structures consisting of communities of genetically closely related organisms. But there are many other properties defining a social order than mere genetic relatedness, things like trade, migration, culture and the like. As the final example of a would-be world in action, let's turn our attention to the world of *Sugarscape,* where these cultural properties are the primary focus of interest.

## Societies in Silicon

The Anasazi, a southwest Native American culture that flourished over 800 years ago, has been a source of deep mystery for archaeologists. How is it, they ask, that a seemingly thriving civilization like the Anasazi could have disappeared almost overnight? Even though there are masses of data available on weather patterns, crop yields, and other environmental conditions, the answer to this puzzle has eluded investigators, mostly because it is impossible to go back and "run the experiment" over again, changing one or two of these myriad factors to see what effect, if any, it might have had on the migratory behavior of the Anasazi. This is the basic dilemma of anyone studying social phenomena: History is an experiment that is only run once. Or is it? Can we really conceive of creating conditions in our computers whereby we could determine what would have happened, say, if the Confederacy had won the American Civil War or if the price of crude oil had only been doubled in 1973 instead of quadrupled? Joshua Epstein and Rob Axtell, two researchers at The Brookings Institution in Washington, D.C., think we can.

In order to conduct the kinds of repeatable, controlled experiments that natural scientists take for granted when trying to understand and create theories of physical and engineering systems, Epstein and Axtell decided to "grow" a social order from scratch by creating an ever-changing environment and a set of agents who interact with each other and the environment in accordance with simple rules of survival. An entire social structure—trade, economy, culture—then evolves from the interactions of the agents. As Epstein remarks about social problems, "You don't solve it, you evolve it." Epstein and Axtell call their laboratory in which societies evolve the *CompuTerrarium*. Here's how it works.

The interacting agents are each graphically represented by a single colored dot on the landscape they inhabit, which is called the *Sugarscape*. Every location in the landscape contains time-varying concentrations of a food resource, called sugar. Each individual has a unique set of characteristics; some are fixed like sex, visual range for food detection, and metabolic rate, whereas others are variable like health, marital status, and wealth. The behavior of these agents is determined by a set of extremely simple rules that constitute nothing more than commonsense rules for survival and reproduction. A typical set of rules might be:

1. Find the nearest location containing sugar. Go there, eat as much as necessary to maintain your metabolism, and save the rest.
2. Breed if you have accumulated enough energy and other resources.

3. Maintain your current cultural identity (set of characteristics) unless you see that you are surrounded by many agents of different
types ("tribes"). If you are, change your characteristics and/or preferences to fit in with your neighbors.

With even such primitive rules as this, strange and wondrous things
begin to happen. A typical scenario is shown below in Figure 4.19, where
we see the sugar marked by yellow dots on the Sugarscape. The agents
are initially distributed randomly on the landscape, red dots being agents
that have a good ability to see food at a distance, blue dots representing
more myopic agents. It is reasonable to expect that if no other considerations enter, natural selection would tend to favor good vision over time.
Indeed this is the case, as seen by the center panel in the figure, showing a preponderance of red agents in the population. However, if the
experiment is run again, giving agents the possibility of passing wealth
on to their offspring in the form of sugar, we find that inheritance has
a pronounced effect on survival. This is shown in the third panel of the
figure, in which many more agents having poor vision are able to survive
by making use of sugar willed to them by their parents.

Although this simple example is useful in illustrating the workings of the *CompuTerrarium,* it hardly suggests a revolution in our way
of thinking about and studying social structures. For that we need to
add a lot more whistles and bells to the system. Epstein and Axtell
have done exactly this. When they add seasons so that sugar concentrations change periodically over time, the agents begin to migrate. When
a second resource, spice, is introduced, a primitive economy emerges
as a result of the new elementary rule: "Look around for a neighbor
having a commodity you need. Bargain with that neighbor until you

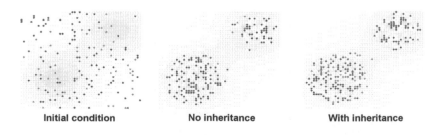

**Initial condition**     **No inheritance**     **With inheritance**

**Figure 4.19**  Evolution on the *Sugarscape* (see Plate XI).

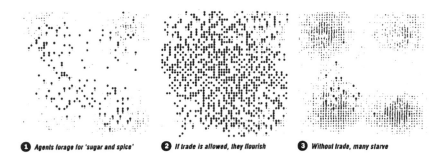

**❶** *Agents forage for 'sugar and spice'*   **❷** *If trade is allowed, they flourish*   **❸** *Without trade, many starve*

**Figure 4.20**   The effects of trade in the *Sugarscape.*

reach a mutually satisfactory price. Trade at that price." Figure 4.20 shows the effect of this type of trading economy. In the first part of the figure, agents are simply foraging independently for both sugar and spice. In the middle panel we see the effect of beginning trade: Now lots of agents flourish. Finally, the third panel shows the effect of turning trade off. Without trade being allowed, many of the agents cannot survive. Quite surprisingly, even this primitive exchange-type of economy provides us with enough structure to construct an experiment to test a version of one of the most cherished theories in all economics: the so-called *efficient market hypothesis.*

### Efficient Markets

According to classical economic thought as taught in the textbooks, in an economy in which all agents have unchanging preferences and live long enough to be able to process all available information about the supply and demand of goods within the economy, prices should settle into a fixed, equilibrium level for all goods. In other words, any price imbalances in the economy will quickly be arbitraged away. This is one version of the efficient market hypothesis, which asserts that all available information is instantly assimilated into the prices of goods in the economy. If there is a shortage of, say, sugar, this fact should get around with the speed of light and the price of sugar should immediately shoot up. We can use the *CompuTerrarium* to test this hypothesis.

The first part of the test involves giving agents the characteristics attributed to them by conventional economic wisdom, namely, unvarying individual preferences and infinite life spans for processing information.

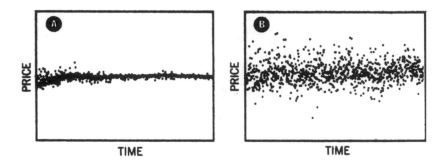

**Figure 4.21**    Price movements in a pure exchange economy.

Classical theory then predicts that prices for goods will settle down to
a single, fixed price at which sugar and spice will be traded, and this
is indeed the case, as shown part (a) of Figure 4.21. The second half
of the experiment involves resetting the system to its original state, and
endowing the agents with *human* qualities: finite lifetimes, the ability
to change preferences, and the opportunity to produce children whose
preferences may differ from their own. Under these far more realistic
assumptions, prices move about in the more-or-less random fashion seen
in part (b) of the figure.

In the second, more realistic, version of the economy, the trading
price never settles down to a single, uniform level but rather seems to
move randomly—but not totally. Although the random price fluctuations
seem to continue indefinitely, it appears as if these fluctuations are all
movements around a single price level. That level is exactly the equilibrium level attained under the idealized assumptions underpinning the efficient market hypothesis. This strongly suggests that the efficient market
version is simply a limiting case of more realistic versions, each of which
corresponds to a weakening of the La-La-Land assumptions of classical
economic theory. Presumably, the more realistic the assumptions, the
greater the variation in price movement about the efficient-market level.
As Epstein remarks, "If the agents aren't textbook agents—if they look
a little bit human—there is no reason to assume markets will perform
the way economic textbooks tell us they should." It's not just in the area
of efficient markets that the *CompuTerrarium* allows us to test theories
of socioeconomic behavior.

Another basic question of economic life is whether, under a given
set of environmental conditions, the overall wealth distribution in society will be approximately equal, or will the agents self-organize into a

so-called Pareto distribution, in which a few of the agents control a disproportionately large share of the overall wealth? Clearly, which of these possibilities will occur depends on the particular trading rules the agents use, as well as the rules governing inheritance, migration, and resource regeneration. Experiments with a wide variety of different choices for these parameters turned up the possibly surprising result that it is very difficult to find conditions under which the wealth of a society ends up being evenly distributed. This would suggest that there are sound theoretical reasons for why various egalitarian socialist ideals are as difficult to achieve in practice as they seem to be: They just do not accord very well with basic human nature, at least insofar as that nature is embodied in rules of economic interaction.

There is certainly much more that can be said about the social laboratory constructed by Epstein and Axtell. Issues involving the emergence of cultural groups, combat, institutional structures, and the like can all be introduced to study myriad questions of interest to social scientists. The interested reader will certainly want to consult the monograph by Epstein and Axtell that details these and many other matters. It is cited in the references for this chapter. Here we must content ourselves with simply noting that the *CompuTerrarium* offers a platform to study society from the bottom up. With this view, we can explore social behavior that is dynamic, evolutionary, and locally simple. What could be better than to have a laboratory like this in which to do such experiments?

## Judgment Calls

In the opening chapter we considered several properties tending to separate good models from bad. These included

- *Fidelity:* the model's ability to represent reality to the degree necessary to answer the questions for which the model was constructed;
- *Simplicity:* the level of compactness of the model, in terms of how "small" the model is in things like number of variables, complexity of interconnections among subsystems, and number of *ad hoc* hypotheses assumed;
- *Clarity:* the ease with which one can understand the model and the predictions/explanations it offers for real-world phenomena;
- *Bias-free:* the degree to which the model is free of prejudices of the

modeler having nothing to do with the purported focus or purpose of the model;

- *Tractability:* the level of computing resources needed to obtain the predictions and/or explanations proffered by the model.

To make these notions concrete, let's evaluate some of the would-be worlds presented in this chapter on these different scales.

## TRANSIMS

When it comes to fidelity, *TRANSIMS* is about as faithful a model of its real-world correlate as any would-be world discussed in this book—or any such world that I know about, for that matter. The details of the Albuquerque road network, the demographics of the travelers in the system, and the fine-grained resolution of traffic movement are represented to an almost unprecedented degree of accuracy. So high marks for *TRANSIMS* along the fidelity axis.

Unfortunately, the *TRANSIMS* model of the Albuquerque road-traffic network is very far from being simple, at least in the sense that it consists of several hundred thousand variables (agents). However, the general mode of behavior of the individual travelers in the network, along with their local interactions with each other are quite simple and straightforward ("go straight, turn left or right, stop," etc.). So on the criterion of simplicity, *TRANSIMS* rates a measure of something like "medium-simple."

As for clarity, *TRANSIMS* certainly qualifies for high marks. The graphical output of the model makes it transparently clear how traffic patterns, congestion, flow densities, and the like appear and disappear. Moreover, the model provides very detailed information about any individual traveler in the system, and the kinds of rules that traveler is using at any given moment. So the clarity of the model is excellent.

Things are a little less fine with *TRANSIMS* when it comes to tractability. While there is no theoretical barrier in the model to carrying out the requisite computations, the sheer volume of calculations required, both to update the state of every traveler and to display the information, is enormous. And, in fact, the full-scale version of the model requires the computing power of a Connection Machine 5 for its execution. This is not exactly computational intractability—but it's far from the type of computational power that is routinely available to all potentially interested users.

Finally, *TRANSIMS,* like the other models in this section, appears to be quite free of investigator bias. As far as I know, neither Chris Barrett's personal feelings about the Albuquerque city government nor his political affiliation(s) nor the status of his bank account enter into any of the assumptions or structure of *TRANSIMS.* All that's programmed in to the model are simple rules about how travelers behave when they drive their cars under given sets of circumstances; everything else emerges as a consequence of these rules of behavior and interaction. So, again, high marks for *TRANSIMS* for being bias-free.

### Tierra

According to its developer, Tom Ray, *Tierra* was never intended to be a nitty-gritty, down-to-details model of real-world biological evolutionary processes. Rather, its goal is to serve as a *metaphor* for evolution. And the way it does this is to pick out a handful of the most characteristic properties of how populations of genes evolve—things like mutation, crossover, and selection. *Tierra* then attempts to mimic these functional activities in strings of computer code, ignoring all the myriad detailed features of real biological genes and their ways of changing over the course of time. So the fidelity between genes in *Tierra* and real-world genes is middling, at best.

The abstractions Tom Ray makes to create computer genes leads to a model that is very simple and clear, in the sense that the number of variables is small relative to the insight the model gives into the work- ings of the evolutionary process. Of course, there are many parameters one can change in the model to create different types of both genetic structures and the environment to which the genes must adapt. But, on balance, *Tierra* certainly stands up to scrutiny rather well when it comes to both simplicity of structure and clarity of results.

Similar remarks apply to the computational tractability of *Tierra.* Even though the model has been implemented on a spectrum of compu- tational platforms, ranging from simple PCs to high-end workstations— and even on distributed systems over the Internet—most of the features of evolution that *Tierra* can help us understand can be found on the low-end systems that almost anyone can access. Moreover, the computational work needed to actually run experiments is modest enough that it's not necessary to wait days or months to get interesting results. A few minutes—or hours on a low-end machine—is all that's required.

The only investigator bias that I've noticed in the *Tierra* setup is Tom Ray's professional prejudices about what is and isn't important in the evolution of genetic populations. And rightly so. But whatever Ray may think about the social order, the world economy, or the political arena does not seem to have found its way into the workings of *Tierra*. So, as with *TRANSIMS,* we find a bias-free model in *Tierra.*

### Sugarscape

Josh Epstein and Rob Axtell's *Sugarscape* model of the evolution of cultural phenomena is even more of a metaphor than *Tierra* is. While I suppose we all have a bag of sugar and a jar of spice somewhere in our kitchens, it's a bit of a stretch to imagine that we spend our working day acquiring the resources needed to stock our shelves with these very items. They are simply stand-ins for different types of things that we need and/or want, as are the sugar and spice mountains and valleys of the *Sugarscape* playing field. So the literal fidelity of the model is very low indeed. This fact raises questions about the degree to which one can take the evolution of economic rules of interaction, "international" trade, and the like in *Sugarscape* as indicative of what and how similar things occur in the real world. On the other hand, the model shows that such behaviors can emerge from very simple rules of interaction among very primitive economic agents. So if you're in the business of trying to model economic processes, the *Sugarscape* results strongly suggest that a good starting point might be to consider these simple rules and see where they lead.

The low fidelity of the *Sugarscape* model, however, leads to a high degree of simplicity, as the agents have a very limited repertoire of rules and interactions available to them as they try to satisfy their sugar-and-spice needs. This is a great asset in understanding what's going on and why the agents act in the way that they do. Moreover, the clarity of the goings-on in *Sugarscape* is very high, since the behavior of the agents is very easy to represent pictorially in the two-dimensional world of the *Sugarscape.*

Equally high marks go to *Sugarscape* for computational tractability. Just as with *Tierra,* the model is designed to run with low-end computational resources of the sort found on personal computers like typical PCs and Macs. So experiments on the *Sugarscape* can be carried out in a few minutes by just about anyone.

We have now looked at three would-be worlds—*TRANSIMS, Sugarscape, Tierra*—and seen how each in its own way sheds light on a real-world system. The one thing we have said nothing about, however, is how one actually goes about creating such a world. Let's close this chapter with an account of some of the nuts-and-bolts aspects of this question.

## Building Silicon Worlds

Each March, the Santa Fe Institute holds a one-day symposium at which members of the SFI research staff talk about their work before the Institute's Science Board. In 1994, Melanie Mitchell presented some of her work on adaptive computation and learning. During the course of this presentation, she gave a checklist of the typical steps one follows in constructing an exploratory simulation of the type we have been considering in this chapter. Mitchell's list is about as good an account as one can find of what must be done to put together a world like *Tierra* or *TRANSIMS,* so let me summarize her account here. Creation of the worlds according to Mitchell follow this sequence:

- Simplify the real-world problem as much as possible, keeping only what appears essential to answering the questions being asked.

- Write a program that simulates the individual agents of the system, their individual rules for action and interaction, along with whatever random elements appear to be needed.

- Run the program many times with different seeds for the random-number generator, and collect data and statistics from the various runs.

- Try to understand how the simple rules used by the individual agents give rise to the observed, global behavior of the system.

- Change parameters in the system to identify sources of behavior and to pin down the effects of different parameter settings.

- Simplify the simulation even further, if possible, or add new elements that appear to be needed.

By and large, these steps are the same ones that an investigator would follow in creating a mathematical, as opposed to a computational, model of a real-world phenomenon. However, there is one big difference: the step involving the writing of the program. The computer-literate reader will have already noted that the chore of putting together all the data elements representing the individual agents and their rules of action and interaction, as well as coding the environment in which the agents interact, is generally a very tricky and time-consuming task.

There is also the crucial matter of how to test the structure and robustness of such a model. The large parameter spaces and nonlinear interactions characteristic of these models make understanding their behavior via standard statistical methods difficult. Recently, John Miller of Carnegie Mellon University has proposed a procedure using genetic algorithms to efficiently explore the parameter space looking for circumstances that would "break" the model's implications. Details of Miller's approach are found in the material cited in the references.

In the worlds discussed here, the individual investigators—Tom Ray, Josh Epstein, Rob Axtell, Chris Barrett—and their collaborators had to carry out these software engineering chores themselves on an as-needed basis, essentially reinventing the wheel with each different world. Noting the inefficiency of this duplication of effort, Chris Langton decided a couple of years ago to take time off from his research on artificial life, in order to develop a general-purpose simulation tool to make it easy to build simulation models. He calls it *SWARM*. Although there is no room here to give a full account of this tool, it is central enough to our concerns to spend a page or two describing the philosophy and the content of *SWARM* in broad outline. Let me also note that this simulation system is publicly available. Details on how to obtain it can be found in the references for this chapter.

### The SWARM Simulation System

The *SWARM* simulation system aims to provide researchers with a standardized, flexible, reliable set of software tools for experimenting with complex, adaptive systems of the type we have been talking about in this book. In particular, *SWARM* offers general-purpose utilities for designing, implementing, running, and analyzing these multiagent systems.

The principal goal of the *SWARM* system is to relieve researchers of the burden of having to deal with computer-science issues arising in the construction of large-scale computer simulations. To accomplish this task, *SWARM* offers a wide spectrum of tools, together with a kernel that drives the simulation. Researchers are then free to customize the general-purpose objects in *SWARM* to model systems in a given area of interest.

The worlds one can create with *SWARM* may differ greatly in their properties, ranging from two-dimensional planar worlds like that of *TRANSIMS,* in which mobile agents move about physically, to abstract graphs, representing communication networks through which agents trade things like messages or commodities. Whatever the universe of discourse, *SWARM* provides a general, uniform framework that allows workers to concentrate on the specific system of interest without being encumbered by issues of data handling, user interfaces, and other purely software-engineering and programming matters.

To illustrate, suppose we are interested in simulating a colony of ants using the *SWARM* system. The first task would be to select a generic two-dimensional region from the *SWARM* library. This will represent the physical region in which the ants from the colony move about. We next choose a generic class of agent from the library, an agent that already knows how to move around in this world, how to sense the state of this world, and how to communicate with other such agents. Once this generic world and agent have been selected, we specialize these generic modules by adding specific ant-world attributes and rules to them. For example, we would probably want to include features for storing the location of the ant nest, tracking the concentration of pheromone throughout the world, and a mechanism for diffusing the pheromones over the course of time. These ant agents would then inherit two-dimensional movement code from the original generic agent. Our next step is to add special code to the agents so that they can sense pheromone concentrations and gradients, as well as adding code for enacting ant rules that decide where the ant will move next. This specialized space and ant agents is then combined in the *SWARM* system with a schedule of activity, which finally gives us a custom-built simulation of ant behavior. Readers interested in the actual implementation details of the *SWARM* system should consult the discussion given in the references.

In this chapter we have looked at would-be worlds ranging from the microsimulation of real urban road traffic of *TRANSSIM* to the metaphorical study of general evolutionary processes provided by *Tierra*. Between these two extremes—one capturing as much structural and real-world detail as possible, the other focusing on the functional features of the process—we found the *CompuTerrarium*, which combines features of both. With these examples of artificial worlds in hand, we can finally return to the questions that engaged our attention in the opening chapter: What can we really expect to learn about the real world of natural and human affairs through the medium of these electronic surrogates?

# CHAPTER

# 5

# Reality of the Virtual

## Inside Information

Aside from serving as the cinema debut of Raquel Welch, the 1966 film *Fantastic Voyage* was notable for its stunning special effects depicting what the inside of the human body might look like if you were a being the size of a blood cell moving about inside it. Interesting as the film was from the standpoint of speculation about how the human body might look from the inside, this experiment in human nanotechnology missed a golden opportunity to ask what today we might term the fundamental question of endophysics: Do the laws governing the behavior of a system look different for an observer standing *inside* the system (e.g., the human body) than for an observer looking at the system from the outside? Put more prosaically, would Raquel Welch, in her form-fitting white jumpsuit, see a different law governing the flow of electrical impulses in the brain than her full-sized human counterpart observing those same impulses from the outside with, say, a microtip electrode?

There is a fundamental question to be settled here between physics from within and physics from without, a distinction that reflects one of the most basic dichotomies in nature and in life: the difference between being inside and outside something—a building, a country, a society. This very same issue lies at the heart of what can be learned from the would-be worlds considered in this book. So let's hammer home the basic problem by considering a few examples of the endo/exo distinction as it arises in science.

**Relativity Theory.**   When asked how he came to discover the theory of relativity, Einstein replied that he imagined how the world would look if he were riding on a beam of light. Here we have a perfect example of endophysical thinking, which in this case led Einstein to the startling conclusion that physics would indeed look different if you were on a light beam than if you were merely watching it flash by. Specifically, you would see yourself traveling at a definite, finite velocity, which in turn would engender all the by-now familiar time-and-space contractions in other systems observed from *outside* the vantage point of your beam of light. But from the light beam itself, you see nothing unusual—just normal speed-of-light travel from one location to another.

**Gödel's Theorem.**   The key message of Gödel's famous Incompleteness Theorem is that *within* any consistent system of logic, powerful enough to express any statement about whole numbers, there are statements that can be made—but which cannot be proved or disproved using the rules of that particular logical system. Nevertheless, by jumping outside the system ("jootsing") we can see that such statements are actually true. They just cannot be proved using the rules of inference contained in the given logical framework. From an endophysics point of view, what Gödel's result says is that arithmetic is incomplete—when looked at from the inside. However, it can be completed if we look at the same system from the outside—exophysically, so to speak. Thus, the laws of arithmetic do look different, depending on whether you look at them from inside or outside a particular logical system.

**The Genetic Code.**   One of the seminal achievements of modern molecular biology is the unraveling of the genetic code by Francis Crick and Sydney Brenner in the late 1950s. This code translates three-letter combinations of the nucleotide bases A(denine), G(uanine), C(ytosine) and T(hymine) into the 20 amino acids that make up the proteins forming all known living objects. The translation process begins with the passive copying of a fragment of the DNA chain into a strand of messenger RNA (mRNA). This mRNA is then read by a ribosome that moves along the strand, one triplet of bases at a time, much like a tape-recorder head traveling along a reel of tape. As it moves from one end of the mRNA strand to the other reading the symbols written on the strand of mRNA, the ribosome captures the amino acid called for by the triplet from pieces of transfer RNA floating about in the cellular cytoplasm. This, then, is a statement of the processes of genetic transcription and translation in a nutshell.

Following Einstein's lead, let us think endophysically about this situation. Suppose we hitch a ride on the ribosome as it passes from one end of the mRNA strand to the other. What would we see? Well, not much. More precisely, what we certainly would *not* see is any clear-cut distinction between the purely syntactical operation of passively copying information from the DNA in the transcription phase and semantically meaningful decoding carried out in the translation mode, in which the syntactic information is used to build proteins. In both cases, all that would be seen from a "molecule's-eye view" would be purely chemical operations and transformations. Only by "jootsing" can we see the syntactic/semantic distinction between the two operations. From the inside, there is only biochemistry; from the outside, there is a whole universe of meaningful proteins.

**Behavioral and Cognitive Psychology.** In the 1920s, John Watson made the radical suggestion that human behavior patterns do not have mental causes. Put more precisely, Watson's assertion was that it is unscientific to claim that there are unmeasurable mental states of the brain giving rise to observable behavior. Thus arose the school of behavioral psychology, which focuses on externally observed input/output or stimulus/response behavior patterns as the raw material upon which theories of the mind and behavior should be based.

By way of contrast, cognitive psychologists argue that it is only by assuming the existence of *internal* states of the brain (i.e., configurations of the neurons) that we have any hope of constructing a genuine scientific theory of human behavior. The cognitive view asserts that it is the job of the psychologist to infer somehow the nature of these internal states from observed behavior, and then to construct a predictive theory of human behavior on the basis of the relationships linking these postulated internal states of the brain.

From an endophysical standpoint, we see the hardware of the brain reconfiguring itself into different states as neurons flash on and off. These states, in turn, give rise to thoughts and behaviors. But an exophysical perspective would show no such states upon which to base the aws of thought. This distinction mirrors perfectly the competing claims in artificial intelligence (AI), in which the pro-AI forces argue what, in effect, is the exophysical position: If it behaves like a brain, then it *is* a brain. Anti-AI proponents reply that you cannot judge a brain by its cover. Pursuit of these competing claims has been well chronicled elsewhere, so let me refer the reader to the excellent treatments cited in the references.

The foregoing distinction between the interior and exterior appearances of the system mirrors rather well the difference between what we have been calling a simulation versus a model. Suppose we have two sorts of objects, let's say a Lamborghini Diablo VT sports car and a second object that someone claims is a duplicate, or a model, of the Diablo. Just what would this mean? What would it take to be a model of a Diablo? It means just what any 10-year-old kid interested in cars thinks it means, namely, that there is a direct correspondence between the external stimuli (road conditions, tire pressures, gas-pedal level), internal states (engine revolutions, spark timing, carburetor setting), and behavior (velocity, motor torque) of the Diablo and the external stimuli, internal states, and behavior of the model. However, the correspondence need not necessarily be either one-to-one or complete, so there may be some external stimuli, states, and/or behaviors of the real Lamborghini that are not represented in the model.

So, for example, when you go to the Lamborghini factory outside Bologna, Italy and look at a model of a Diablo in the designer's office, the speedometer, door handles, spare tire, and all the other paraphernalia whose configurations form the internal states of the real Diablo are not present in the model. And they are not there for the very good reason that they are irrelevant to the model's purposes, which center primarily on showing the overall design and structure of the car. Nevertheless, the external stimuli, states, and behaviors of the model are in direct relationship to a subset of the external stimuli, states, and behaviors of the real car. Such a correspondence generates what we might call a modeling relation between the real Diablo and the object in the designer's studio. Note carefully that the model is *simpler* than the real-world object it models, in the sense that the model has fewer states. This property is characteristic of modeling relationships: Models are always simpler than what they model. So in this sense we can say that the model Diablo is an exophysical description of the real car. Now, what about a simulation?

In my study at home I have a brand-X laser printer whose operating instructions assure me that by suitable fiddling I can make it "emulate," that is, simulate, a different type of printer, a Hewlett-Packard Laser-Jet Plus. What does it mean to say my brand-X machine can simulate another machine? It means simply that the inputs and states of the HP machine can be coded into the states of my printer, and those states of my brand-X printer can be decoded into the appropriate outputs that would be generated by an actual HP printer. Note that in order for such an encoding/decoding dictionary to be set up, my printer must be more complicated than the HP machine in a very definite sense. Specifically, in order for the inputs and states of the HP machine to be encoded into the states of my "simulator," my machine must have more states than the HP printer when both are regarded as abstract machines. Thus, the simulator (my printer) must be more complicated than the object being simulated (the HP printer). This situation is completely general: A simulation is always more complex than the system it simulates in the sense that it contains more states. From an endo/exo perspective, this simulator constitutes an endophysical description of the HP printer, because it contains every detail of the real-world system *inside* another, strictly larger, system.

There is another way of looking at this distinction between a model and a simulation, one that emphasizes the difference between an actual world of experience and a *theory* of that experience. This idea, originally put forth by statistician David Lane of the University of Modena and the Santa Fe Institute, again rests upon the endo/exo dichotomy—but in interesting, new, and illuminating ways. Let us briefly look at what Lane has in mind.

## The Moral of the Model

The *TRANSIMS* world, with which we opened the last chapter, is an example of a simulation *par excellence*. It is designed to generate road-traffic patterns that are comparable in every respect to what might be seen on the real highways and byways of Albuquerque. Not only do the developers of *TRANSIMS* employ exactly the same vocabulary—households, travelers, freeways, traffic lights—as the Albuquerque traffic-system managers, the system is explicitly geared to providing predictions of exactly the observables that these managers care about.

By way of contrast, a would-be world like Tom Ray's *Tierra* is not like this at all. The genomes constituting this world are highly abstracted versions of those found in real-world organisms. These abstractions are based on what David Lane calls "theories about what 'worlds of experience' are like." These theories are based on what kinds of objects (like stock traders) populate such a world, how the objects relate to each other (through, for instance, buying and selling equities), and what kinds of processes change the character of objects, including the creation and destruction of the objects (such as new trading strategies or bankruptcies).

In this view, the relevance of worlds like *Tierra* to their real-world counterparts is mediated by these theories. In other words, the theories constrain the ways we can interpret the models. However, it is very difficult to gain access to the theories using everyday language. Any attempt to describe these theories in ordinary language is rife with confusion and ambiguity. What to do?

According to Lane, probably the best way to gain access to the theories about what the world of experience is like is via the would-be worlds themselves. This is so even though the meaning(s) of these surrogate worlds depends on the theories we are trying to understand. If at first hearing this sounds circular, it is! The circularity of this argument comes from the fact that the meaning in the real world of a world like *Tierra* is derived from the theories, whose meaning itself is perceived through the world. The basic point to observe here is that a model like *Tierra* or the *CompuTerrarium* is designed to highlight specific features like evolutionary adaptation or the emergence of rules of economic behavior. As a result, it is very easy to identify these aspects in the would-be world, and thus it is easy to appreciate their significance *in that world*. Once we understand these properties and behaviors in the electronic context, we can begin to see them in our real world of experience via a kind of reinterpretation of what has been discovered in the surrogate world. With these rather general notions in mind, let's briefly examine what Lane calls "aphorisms" or morals that can be drawn from various models of the *Tierra* type.

In his analysis of the "back-and-forth" causation between theories of the world and the models we construct, David Lane identified four guiding principles that can be drawn about the behavior of real complex systems from their electronic counterparts.

## Chance as a Cause

A few years ago, Lane and Brian Arthur cooked up a simple model of consumer choice to study the process they termed "information contagion." The general structure went like this. Suppose two competing products enter the market, for example, two types of personal computers like the Apple Macintosh and the IBM PC. For each of these products, a certain amount of readily available advertising from the manufacturers is available to help consumers decide which machine to buy. Purchasers also make use of other information, such as hearsay evidence from previous buyers of these machines, reviews in computer magazines, and even free trials offered by computer vendors. On the basis of this information, a new customer buys either an Apple or a PC machine, thereby entering the pool of previous purchasers that will later provide information that newcomers will sample.

Lane and Arthur wanted to discover what kinds of market structures would emerge from the foregoing type of information dynamics. They found that typically only a very few types of stable market structures were possible. The types that could occur depended on the actual performance characteristics of the competing products, as well as on how buyers used the information they obtained in arriving at their decision about which item to purchase.

In the special, although not uncommon, case in which the products actually have the *same* performance characteristics (a fact that is, of course, unknown to the prospective consumers), it turns out that there are only three possible types of market structures: the two products either split the market 50–50, or one of the two products completely dominates the market. The overriding question here is which of these three possibilities will actually take place.

The short answer to this question is that it is completely unanswerable. There is no way to know *in advance* whether the products will split the market or if one product will dominate. The reason for this undecidability is that the overall market pattern is constructed from the many samples, and hence choices, made by individual buyers of computers. Because this sampling is completely random, the aggregate market structure itself is randomly determined as well. Thus, it seems appropriate to say that chance is the cause of the structure that the market ultimately assumes. Note that chance as a cause does not mean that *anything* can happen; the forces of chance sampling are constrained by the global fact that the system can end up in only one of the three types of aggregate market structures.

As Lane points out, the issue raised by this model of market behavior is not what we don't know, but rather what cannot be known. Using a different example, how do phenomena, like the actions of an individual driver that may operate below our level of observation, generate structure at the levels at which our observations and actions actually do take place? This behavior is about as good a description of what we mean by emergent behavior as any.

Remarkably, our everyday notion of causality does not deal with this kind of interaction between levels of experience very well, at all. For instance, if we try to explain something like the collapse of the savings-and-loan industry in the United States, the most common mode of explanation is simply to say that if something fails, then somebody did something wrong. Thus, the remedy is to find out who did what and get rid of them. This is exactly what was done in the savings-and-loan fiasco, when people like Charles Keating were sent up the river for a few decades, thus serving as scapegoats for the entire mess. In this type of explanation for a phenomenon, the presumed cause resides at the level of individual action. However, there is another way to explain things.

The second mode of explanation is to appeal to some law operating at the aggregate level, in this case the level of the savings-and-loan industry itself. Such a law might be something like, "It is inevitable that when money becomes so plentiful and regulators look the other way, unscrupulous bank managers will help themselves to a share of the loot." Neither of these explanatory modes—the individual or the aggregate—look for causes in the relation between the interactions of single decision makers, like Keating at the individual level and the emergent patterns such as the collapse of the entire savings-and-loan industry at the aggregate level. Thus, neither mode can accommodate a causal role for chance.

## Winning Isn't Necessarily Winning

In the world of *Tierra,* we saw different strategies for survival come and go as time went on. When the soup was inoculated with the Ancestor organism, an ecology of strategies (genotypes) developed, which dominated the *Tierra* landscape for many generations. Generally speaking, these ecologies consist of a rather small set of strategies, existing in the population in some fixed proportion to each other. Often, no one genotype dominates *Tierra,* because there may be various cycles of predation (A eats B who eats C who eats A), parasitism (A freeloads off B but B does not take from A, at all), mutualism (A and B help each other).

Sooner or later, however, such an ecology of strategies fails to respond well to a mutant. The mutant strategy gets a foothold, stability of the ecology is lost, and some transient period ensues in which the entire population restructures itself. After this transient phase, the old ecology is replaced by a new one. This is the phenomenon that evolutionary theorists call "punctuated equilibrium." Let's look closer at this successful mutant strategy that causes all the fuss.

Typically, the new strategy that sets in motion the restructuring of the original ecology disappears as the old ecology is torn down. Moreover, after all the smoke clears away and the new ecology is set in place, this strategy does *not* reappear in the successor ecology. The strategy's success is generally due to its ability to compete successfully with one of the old dominant types. But when the frequency of this type of strategy is sufficiently reduced as the old ecology starts to crumble, the invading strategy finds that its way of playing no longer works against the new strategies that emerge.

It is important to observe here that this phenomenon of "winning isn't necessarily winning" is quite different from the old saw, "he won the battle but lost the war." That piece of folk wisdom implies that a temporary advantage was exploited for a short-term gain, but that this gain may cost more to the victor than it does to the loser. In our situation, the whole notion of an adversary may be meaningless, because the relationship between a given strategy and any other may range from complete competition (predator–prey) to complete alliance (symbiosis).

So in a coevolutionary world like *Tierra,* an agent that defines its success by winning against a currently dominant strategy may find itself becoming the victim of its own success. David Lane speculates on what it might mean for such an agent to *have* a strategy rather than *being* a strategy. Such an agent would have a set of rules for choosing its actions in response to what has taken place, rather than just *being* a rule as it is in the world of *Tierra.* How might such an agent predict the onset of destabilizing periods, and recognize when a new ecology was beginning to form? What types of actions could such an agent take to ensure that it became part of a stable network of relationships with the newly emerging agents? Current strategic-management practice does not ask these types of questions. But it is hard to imagine that they would not be of interest in today's fast-paced business world of mergers, acquisitions, and startups, not to mention the rapidly shifting allegiances in the political arena.

## Organization as Structure and Process

Walter Fontana is a self-effacing, soft-spoken, bespectacled Südtiroler who does research in biochemistry at the University of Vienna and the Santa Fe Institute; he is also a man with a mission. That mission is to understand how collections of objects—molecules, organisms, economic agents—interact to form organizations. When it comes to thinking about organizations, be they countries, clubs, corporations, or the military, we tend to see them as defined by their structure. This leads to a mental picture of an organization as an elaborate diagram outlining the hierarchical chains of command and control within the organization, chains that link lots of little boxes into which the members of the organization are pigeonholed. However, there is another way to think about an organization: It is a collection of *processes,* which suggests envisioning the organization of an outfit like General Motors or the Roman Catholic Church as being simply a way of accomplishing some task—selling cars or selling salvation, take your pick. Fontana's mission is to try to understand *both* these features of an organization—structure and function—in a single, unified framework.

As a chemist, Fontana was painfully aware that most scientists regard chemical systems as consisting of a collection of passive objects—molecules, typically—that are shoved around by energy operators to create new compounds. This is basically a structural view of chemistry. Taking a functional view, however, makes it far less clear who is shoving who around inside the test tube. Because the basic reaction in chemistry is that object A acts on object B to produce object C, it makes just as much sense to regard both A and B as being active operators, whose interaction creates C as it does to envision them both as passive objects that are somehow pushed together. Furthermore, there may often be an alternate reaction $D + E = C$, involving objects D and E, totally different from A and B, that produces the same final product. In this case, we think of the action of A on B as being equivalent to the action of D on E.

These basic reactions—the rules for determining reaction products and the rules for equivalent objects—constitute what is abstractly called a *lambda calculus* of objects. Fontana's idea was to take a lot of lambda objects, toss them into a pot, and let them interact randomly. After a large number of such collisions, he stopped the system and looked into the lambda soup to see what kinds of objects had been created (and destroyed).

What Fontana found was, at first, not very encouraging. No matter how he tried to vary the initial set of objects and the number of collisions, he always seemed to end up with a collection of objects whose function seemed simply to copy themselves. These self-copiers are the ultimate egoists; they just greedily copy themselves, using up all the resources to do so, thus crowding all the other objects out of the soup. Clearly, such a collection of egoists does not display much by way of interesting organizational structure, because in such a world it is literally "every man (or molecule) for himself."

As a last, desperate attempt to get something interesting to emerge out of this soup, Fontana decided to change the rules of interaction: He simply banned self-copying. Basically, the new rule was that if any reaction produced a product that was one of the reactants, that reaction was declared null and void; it simply did not exist. Now Fontana was getting somewhere. With this constraint in place, all the new objects created belonged to one or more *sets* of objects that Fontana called "self-maintenance." This meant that every object in such a set was the end product of a reaction involving members only of the very same set the object itself belonged to. So the *set* could reproduce itself—even if no single member of the set could do so. As he continued to experiment with such abstract chemical systems, Fontana began to create a whole library of patterns for self-maintaining sets.

At this juncture, Fontana joined forces with biologist Leo Buss of Yale University. Now with biological structures like multicellular organization or ecological foodwebs on their minds, the two of them started thinking of these self-maintaining sets as abstract organizations.

They performed many interesting experiments involving the interaction of elements from two different self-maintaining sets. These experiments showed that in some circumstances one organization would drive the other to extinction. However, under a different set of conditions, it could happen that the two organizations would combine into a higher-level self-maintaining set, much like two corporations in different lines of business—Turner News Network and Time-Warner, for example—merging to form a megacorporation stronger than either taken separately. Interestingly, Fontana and Buss discovered that this kind of corporate merger typically required additional lambda objects, generated by cross-reactions between the two component organizations, to stabilize the new structure.

This line of research is still ongoing, and it's hard to say at the moment exactly what conclusions one can draw from it for our everyday social, economic, and business organizations. But the one point that is made very clear by this work is that it is important to think of organizations as processors, not simply structures, and that the very processes that an organization carries out generates its structure. So it is process that's paramount, not structure.

## Rationality Isn't Necessarily Intelligent

Cosimo de Medici was universally acknowledged by his contemporaries and historians alike as being the central power broker in Florentine politics for over 60 years in the middle of the fifteenth century. According to John Padgett, a University of Chicago researcher who has studied the period, Cosimo attained his unique position through a combination of two factors. The first is that by birth Cosimo was a member of the old aristocratic class. But by the so-called New Men, those with recently acquired wealth and access to power, Cosimo was regarded as their champion. So the question that arises is what exactly was old Cosimo *doing* that enabled him to successfully control the Florentine scene for such an extended period of time?

An explanation of Cosimo's success comes from Padgett's analysis of the time. Briefly, the network of personal and professional interactions that became the Medici power block had a star-shaped configuration, with the Medici at its center. Thus, the other families in the network were connected to each other only through their connection to the Medici.

The Medici consolidated their key role in this structure by marrying into other aristocratic families, as well as by doing business with the New Men. But the New Men with whom they transacted business had no links of any kind with the aristocratic families. Consequently, all of Cosimo's acquaintances could see him through whatever eyes they wished, without fear of contradiction by any of the others. Moreover, any plans for actions on the part of the Medici had to pass through Cosimo before being implemented.

This picture of the Florentine interaction network makes it clear that neither Cosimo nor anyone else actually *designed* the network. Rather, it came about as the end result of what can only be regarded as a series of historical accidents. It is not at all clear how this overall power structure came about from the myriad local interactions among aristocrats and New Men of which it is composed. Nevertheless, a shrewd operator like Cosimo de Medici could sense his positioning in the structure and the opportunities it offered to him.

The historical record gives no indication that Cosimo engaged in any rational planning to make this particular structure come about, a structure for which he was ideally suited to benefit. In fact, it is extremely unlikely that Cosimo could have created this structure had he set out with a coherent, rational plan to do so. Yet Cosimo's general hands-off, low-key style, coupled with his unique position in the network, were mutually supportive, and led to more intelligent decisions and behaviors from the Florentine system than could possibly have resulted from Cosimo following the choices dictated by rational-choice theory. Hence, the conclusion that rationality isn't necessarily intelligent.

So there are four principles, or rules-of-thumb, that can be gleaned from the study of complex systems: chance as a cause, coevolution, organization as process, as well as structure, and limited rationality. The key insight underlying all these nuggets of wisdom is that microlevel interactions between individual agents and global, aggregate-level patterns and behaviors mutually reinforce each other. For instance, rising stock prices encourage individual traders to buy, which in turn acts to push prices still higher. It is the job of the "complexicist" to tease out the logical basis of the chains of causation at work between these very different levels. But these aphorisms and their associated empirical observations are far from constituting what a "hard" scientist would call *laws* of complex behavior. We examined the issue of when an empirical relation like

Zipf's Law for languages can be regarded as a bona fide law of complex behavior. But what about the extreme possibility that no such laws exist? Is there any way we can convincingly argue that there exist phenomena in the natural and/or human worlds that defy explanation by rule?

Throughout this volume, we have been tacitly assuming that these microworld versions of real-world processes like stock markets and ecosystems offer a new tool for answering questions that heretofore have had to remain in the realm of what we earlier termed "hypotheticality." But even if this is the case, perhaps there are still questions that transcend our ability to answer them with this new, souped-up brand of epistemological medicine. After all, we are still using a computing machine to tease out the answers to questions about things like road-traffic flow and the movement of stock prices. So it stands to reason that whatever limits we have on our ability to compute will have direct bearing on the types of questions we can answer even using this newfound tool. Because the work of Gödel and Turing discussed in chapter 3 shows that such unanswerable questions certainly do occur in the world of mathematics, it is well worth spending a few pages looking at the possible limits to our acquisition of knowledge by rule in areas outside the pristine world of pure symbols and logical inference.

## Limits to Scientific Knowledge

To anyone infected with the idea that the human mind is unlimited in its capacity to answer questions about natural and human affairs, a tour of twentieth-century science must be quite a depressing experience. Many of the deepest and most well-chronicled results of science in this century have been statements about what *cannot* be done and what *cannot* be known. Probably the most famous limitative result of this kind is Gödel's Incompleteness Theorem, which tells us that no system of deductive inference is capable of answering all questions about numbers that can be stated using the language of the system. In short, every sufficiently powerful, consistent logical system is incomplete. A few years later, Alan Turing proved an equivalent assertion about computer programs, which states that there is no systematic way of testing a program and its data to say whether or not the program will ever halt when processing that data. More recently, Gregory Chaitin has looked at Gödel's notion of provability from an information-theoretic perspective, finding explicit examples of simple arithmetic propositions whose truth or falsity will never be known by following the deductive rules of any system of logical inference. Essentially, what Chaitin's results show is that such mathematical questions are simply too complex for us. For more details on these pivotal results, the reader can consult the discussion in chapter 3 and the references cited therein.

The theorems of Gödel, Turing, and Chaitin are limitations on our ability to know in the world of mathematics. The same limitation applies to statements such as the celebrated Heisenberg Uncertainty Principle in quantum theory, which at first glance appears to refer to an inherent limitation on our ability to measure certain quantities in the physical world. But a more careful examination shows that Heisenberg uncertainty is actually a limitation imposed by certain *mathematical* formulations of quantum theory, and may or may not be an intrinsic limitation in the structure of the real world itself. Similar remarks apply to limitations like Arrow's Paradox in social-choice theory and the Central Dogma of Molecular Biology (DNA $\rightarrow$ RNA $\rightarrow$ Protein), which are both limitations imposed by *models* of the real world rather than provable limitations about what can be known and/or done in the real world itself. So it is reasonable to wonder if there are questions about the worlds of natural and human phenomena whose answers science is forever powerless to uncover.

The first, and perhaps most vexing, task in confronting this issue is to settle on what we mean by scientific knowledge. Philosophers have grappled for ages with the problem of what constitutes knowledge, with no end to this struggle yet in sight. To cut through this philosophical Gordian knot, let us adopt the perhaps moderately controversial position that a *scientific* answer to a question takes the form of a set of rules or a program. We simply feed the question into the rules, turn the crank of logical deduction, and wait for the answer to appear as the output of program. Of course, not just any set of rules will do. In order for the rules to be scientific, they must pass certain tests of reliability, objectivity, explicitness, public availability, and so forth, as outlined in the opening chapter, but once such filters have been applied, what remains is pretty much an algorithmic notion of what we usually consider to be a scientific theory. It is theories that we use to answer questions. Also, the rules we use may exist in any of three quite-different worlds: the physical, the mathematical, and the computational. As we will see below, it is crucial to keep the distinction between the three uppermost in mind as we invoke the rules to answer any particular question.

Thinking of knowledge, scientific-style, as being generated by what amounts to a computer program opens up the issue of computational intractability. We know that there are puzzles, like the celebrated Traveling Salesman Problem, whose computational difficulty is widely believed to increase exponentially with the size of the problem. Thus, to calculate the minimal-cost tour of the 180 or so world capital cities by a brute-force examination of each of the $180! = 180 \times 179 \times 178 \times \cdots \times 2 \times 1$ possible routings would require a time much greater than the age of the universe with even the fastest of computers. But such a computation is possible, at least, in principle. Challenging as such problems are, they are not our primary concern here. Rather, we focus on those questions for which there exist *no programs at all* for producing an answer. Our interest, then, is in questions that are *logically,* rather than practically, unanswerable. Let us turn to some specific examples of questions in the realm of natural and human affairs that serve to motivate what is involved in identifying such logical barriers to science.

### Limits in Nature

In order to bring the issues of limits of scientific knowledge into sharper focus, let's look at three well-known questions from the areas of physics, biology, and economics. Two such questions were already discussed in chapter 3, namely, the problem of the stability of the solar system and the protein-folding problem. Let us briefly revisit these two questions, the one from classical physics, the other from modern molecular biology.

**Stability of the Solar System.**   One version of this almost eternal question involves $N$ point masses moving in accordance with Newton's laws of gravitational attraction. We then ask if there is some set of initial positions and velocities of the particles so that either (a) at some finite time in the future two (or more) bodies collide, or (b) given a finite time in the future, one (or more) bodies acquires an energy greater than any predetermined amount by that time, and thus eventually flies off out of the system (attains escape velocity). In the special case when $N = 10$, this is a mathematical formulation of the question, Is our solar system stable?

As noted in chapter 3, Jeff Xia of Northwestern University gave a definitive answer to the general question of the stability of such systems by constructing a five-body system for which one of the bodies does indeed gain an arbitrarily large amount of energy after a finite amount of time. But this solves only a *mathematical* version of the real-world problem; what it says about a real five-body system is anyone's guess. But more on this later.

**Protein Folding.** In chapter 2, we discussed the fact that the proteins making up every living organism are all formed as sequences of a large number of amino acids, strung out like beads on a necklace. The instructions for what beads go in which positions are contained in the cellular DNA. But once the beads are put in the right sequence, the protein folds up into a very specific three-dimensional structure that determines its specific function in the organism. A geometric view of this situation was shown in Figures 2.8 and 2.9 for *cytochrome-c,* a very short protein consisting of just 104 amino acids.

The big question of protein folding is this: Given a particular linear sequence of amino acids, what three-dimensional configuration will the sequence fold itself into? It is generally thought that the folded configuration of a protein is its lowest free-energy state, and in nature we see proteins composed of several thousand amino acids folding into their final configuration in just a second or so. Yet when we try to simulate this folding process on a computer, it has been estimated that it would take $10^{127}$ *years* of supercomputer time to find the final folded form for even a very short protein consisting of just 100 amino acids. In fact, Aviezri Fraenkel showed, in 1993, that the mathematical formulation of the protein-folding problem is an "NP-complete" problem, which means that it is computationally "hard" in the same way that the Traveling Salesman Problem is hard. So how does nature do it? This is the computational part of the question of protein folding. Now let's look at a candidate question for being unanswerable, one taken from the world of human affairs.

**Market Efficiency.** One of the pillars upon which the academic theory of finance rests is the idea that financial markets are efficient. Roughly speaking, this means that all information affecting the price of a stock or commodity is immediately processed by the market and incorporated into the current price of the security. One consequence of this type of efficiency is that prices should move in a purely random fashion, discounting the effect of inflation. This, in turn, means that trading schemes based on any publicly available information, like price histories, should be useless; there can be no scheme that performs better than the market as a whole over a significant period of time. Or so goes the theory, in any case.

Actual markets don't seem to pay much attention to academic theory, and the literature is filled with market anomalies, like the small-firms effect, which states that the stock of smaller companies outperforms that of larger ones, and the low price-earnings ratio effect, which asserts that stocks with low price-to-earning ratios perform better than those having high ratios. Both anomalies cast considerable doubt on the idea of market efficiency. To illustrate this point, the chart in Figure 5.1 shows

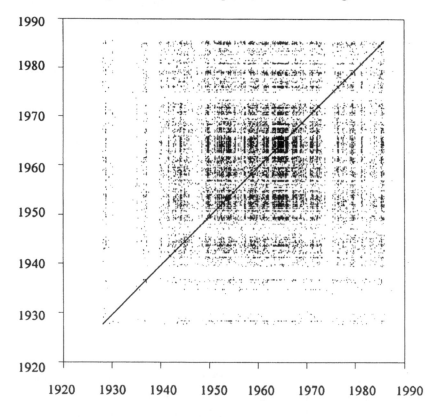

**Figure 5.1**    Recurrence plot for the S&P 500 index, 1928–1985.

a recurrence plot of the Standard & Poor's 500 index over the period 1928–1985. This plot is generated by considering the index at two times, $t$ and $s$, and coloring the point $(t, s)$ in the plane black if the difference in the two values is less than one-half the standard deviation of the entire S&P data set; otherwise, the point remains white. If the market were truly efficient and prices did indeed move randomly, this plot should be shaded uniformly; clearly, it isn't. There are bands of white during the Depression, a turbulent time for stocks, and dark black boxes during the late 1950s and early 1960s, which was a period of very low volatility in the market.

So we have three big questions about the natural world—stability of the solar system, protein folding, and market efficiency—and what appear to be three answers: The solar system may not be stable, protein folding is computationally hard and it is very unlikely that financial markets are efficient. But what each of these putative answers has in common is that it is the answer to a *mathematical question* phrased about a *mathematical representation* of the real-world question, not an answer to the real-world question itself. So, for instance, Xia's solution of the $N$-Body Problem says nothing about *real* planetary bodies moving in accordance with real-world gravitational forces. Similarly, Fraenkel's conclusion that protein folding is computationally hard does not begin to touch the issue of how *real* proteins manage to calculate the right configurations to do their job in seconds rather than zillions of millennia. Finally, canny operators on Wall Street have been thumbing their noses at the efficient market hypothesis for decades.

What these examples show is that if we want to look for scientifically unanswerable questions in the real world, we are going to have to carefully distinguish between the world of natural and human phenomena and mathematical and computational models of those worlds. Unlike conventional scientific investigations, in which it usually does no particular harm to mix these worlds, here the separation of the worlds of nature and mathematics is crucial if we want to avoid throwing out the baby with the bath water. At any given moment, we have to be clear in our minds about whether we are making a mathematical argument about the properties of a mathematical representation of the real world or whether we are we making a real-world assertion about some causal chain of implications in a real-world system. These are obviously not the same thing. It's useful to look just a bit further into this point.

## A World of Worlds

Our concern is with questions about the real world that may be unanswerable. Yet the only realm in which we have tools available for *proving* a question to be logically unanswerable is in mathematics. Moreover, because we have taken a scientific answer to be what amounts to a computer program, the world of computation is also relevant to the determination of unanswerability. Let's quickly examine each of these worlds in turn.

The objects of the real world consist of directly observable quantities like time and position, or quantities like energy that are derived from them. Thus, we consider things like the measured position of planets or the actual observed configuration of a protein. Such observables generally constitute a discrete set of measurements taking their values in some finite set of numbers. Moreover, such measurements are generally uncertain.

In the world of mathematics, on the other hand, we deal with symbolic representations of such real-world observables, where the symbols are often assumed to belong to a continuum in both space and time. Furthermore, the mathematical symbols representing things like position and speed usually have numerical values that belong to number systems such as the integers, the real, or the complex numbers, all systems containing an infinite number of elements. In mathematics the concept of choice for representing and working with uncertainty is randomness and the tools and methods of probability theory.

Finally, we have the world of computation that occupies the curious position of having one foot in the real world of physical devices and one foot in the world of abstract mathematical objects. When we think of computation as the execution of a set of rules, an algorithm, then the process is a purely mathematical one belonging to the world of symbolic objects and their relationships to one another. But if we regard a computation as the process of turning switches ON and OFF in the memory of an actual computing machine, then it is a process firmly rooted in the world of physical observables.

These three worlds—the physical, the mathematical, and the computational—are depicted schematically in Figure 5.2. It is the relationship among these very different universes that must be kept uppermost in mind if there is to be any hope of creating a viable *theory* of the limits to scientific knowledge. We must somehow bring these three worlds into congruence insofar as they express any particular question, like the

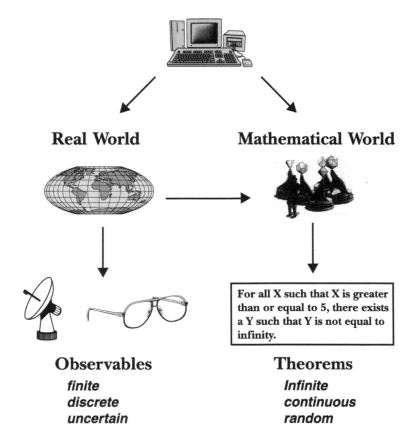

# *THREE WORLDS*

### Computational World

### Real World

### Mathematical World

For all X such that X is greater than or equal to 5, there exists a Y such that Y is not equal to infinity.

### Observables

*finite*
*discrete*
*uncertain*

### Theorems

*Infinite*
*continuous*
*random*

**Figure 5.2**    The worlds of observables, mathematics, and computation.

stability of the solar system. Let us consider a couple of approaches to how this kind of congruence might actually be attained.

### One World or Two?

Let's review the bidding. We are concerned with the existence of questions about the real world that are logically impossible to answer by scientific means. To demonstrate the existence (or nonexistence) of such questions, there are two choices: restrict all discussion and arguments solely to the world of natural phenomena, or mix the worlds of nature and mathematics *cum* computation.

Suppose we choose the first path, remaining exclusively within the confines of the natural world. This means that we are forbidden to translate a question like, "Is the solar system stable?" into a mathematical statement and employ the logical proof mechanism of mathematics to provide an answer. The difficulty here is that our goal is to *prove* that the question is scientifically (un)answerable. But the very notion of a proof exists only in the world of mathematics. Therefore, we face the problem of finding a substitute in the physical world for the concept of proof.

A leading candidate for replacing proof is the notion of causality. Adopting this as a substitute in the world of natural processes for the mathematical notion of proof, we are led to say that a question is scientifically answerable if it is possible to produce a chain of causal arguments whose final link is the answer to the question. For instance, the issue of the stability of the solar system might be settled by a causal chain beginning with the positions, velocities, and masses of the planets, which leads via causal arguments to a plot of the planetary orbits. These, in turn, would then show (causally) that no collisions or escapes are possible up to existing observational accuracy. However, to construct a convincing causal chain in complicated situations, especially those involving human participants, may be a daunting task. So let us consider approaches that mix the worlds of nature and mathematics.

If we want to invoke the proof machinery of mathematics to settle a particular real-world question, it's first necessary to encode the question as a statement in some mathematical formalism, such as a differential equation, a directed graph, or an $n$-person game. We then settle the *mathematical version* of the question using the tools and techniques of this particular corner of the mathematical world, eventually decoding the answer (if there is one) back into real-world terms. The overall process is shown schematically in Figure 5.3. This encoding/decoding process sets up a modeling relation between the real-world phenomena we care about and its mathematical-world representation.

The obvious problem here is the task of providing a convincing argument that the mathematical version of the problem is a *faithful* representation of the question as it arises in the real world. This is simply another way of stating the age-old question of model validity: How do

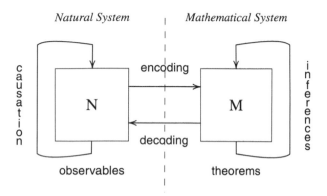

**Figure 5.3**  The modeling relation.

we know that a mathematical model of a natural system and the system itself bear any relationship to each other? This is a *metascientific* question, entailing the development of a theory of models for its resolution. But there is another way.

All existing mathematical results on undecidability rely upon the underlying number system—integers, reals, complex—having an infinite number of elements. But as we noted a moment ago, in the real world of observation, measurements must take their value in a *finite* set. There are no Gödel-type undecidability results for logical systems over these kinds of finite number systems. Similarly, employment of nondeductive modes of reasoning—induction and abduction, for instance—take us beyond the realm of Gödelian undecidability. If we restrict our mathematical formalisms to systems using nondeductive logic and/or finite sets of numbers, then every mathematical question is decidable, that is, answerable; hence, we can expect the decoded real-world counterpart of such mathematical questions to be answerable as well. One approach to the unanswerability question revolves about the employment of this kind of "No-Gödel" formalism.

In a related direction, one might consider whether the human mind is constrained in its creative capacity by being a type of computer in the Turing-machine sense, albeit a very sophisticated one. This, of course, is the central question of "strong AI": Can machines think just like you and me? Recently, studies of this question under the aegis of the Institute for Future Studies in Stockholm by the author, psychologist Margaret Boden, mathematician Donald Saari, economist Åke Andersson, and others suggest strongly that in the creative arts, as well as the natural sciences and mathematics, it seems highly unlikely that human creative capacity is subject to the rigid constraints of a Turing-type of computer. There are other models for computation besides that of Turing, and it may well be that the human mind is some type of super Turing machine, for example, a DNA computer. These matters are currently under investigation. But if the human mind is able to transcend the following of rules in its cognitive activity, then it would also not be subject to Gödelian limits of undecidability.

On this non-Gödelian note, let me conclude this examination of limits to scientific knowledge with a personal speculation about what would be needed for the world of physical phenomena to display the kind of logical undecidability seen in mathematics. Basically, my claim is that for this to happen nature would have to be either inconsistent or incomplete. What do these mathematical notions correspond to in the world of physical phenomena?

Consistency means that there are no true paradoxes in nature. Quantum mechanics notwithstanding, my view is that particles do not move to the left and to the right simultaneously, nor does water run uphill and downhill at the same time. Every time we have encountered what appeared to be such a paradox, as with the redshift of quasars or the seemingly slow rate of expansion of the early universe, subsequent investigation and theory has provided a resolution of it. So I will take it as an axiom of faith that nature is consistent.

Completeness of nature implies that a physical state cannot arise for no reason whatsoever; in short, there is a cause for every effect. Again, I can think of no incontrovertible counterexamples to this claim. So I will take it as a working hypothesis that nature is not only consistent, but also complete.

Putting these two assertions—consistency and completeness—together, I believe it is likely that there are no logical barriers to providing a scientific answer to any question we care to put to nature. Perhaps a tour of twentieth-century science is not so depressing after all!

Our leitmotif throughout this book has been the idea of using the computer as a tool for experimentation. The novelty resides in the relentless march of technology, a march that has finally provided us with computational capabilities allowing us to realistically hope to capture enough of the real world inside our programs to make these experiments meaningful. This raises the interesting question of what science might have been like in the past if such technology had been available to thinkers like Aristotle, Newton, and Gauss. Let's speculate on this theme for a moment.

## Computation *Über* Alles

Of the myriad stratagems I employ to avoid useful work, one of the most pleasurable is to speculate how scientists of earlier eras would have made use of modern computers and what effect it would have had on the science of their times—and ours. For example, I suspect that Newton's geometrically based arguments for particle motion would have remained essentially untouched by the hand of the machine (although Newton may well have used a relational database to trace out the various biblical lineages that seemed to have occupied most of his time). On the other hand, it's likely that Kepler would have discovered much more than his three laws of planetary motion, perhaps even anticipating Newton's equations, had he had access to a Sun or SGI workstation. There's no doubt in my mind that Gauss would have proved the Prime Number Theorem long before de la Vallée Poussin and Hadamard if he had been able to make use of packages like *Mathematica* or *Maple* to study the distribution of primes.

Recently, while surfing the Internet looking for some now long-forgotten item, I stumbled across a hyperlink to work of the great turn-of-the-century Scottish naturalist and polymath d'Arcy Wentworth Thompson. This serendipitous event underscored again the great advantage of search engines that are not *too* accurate or *too* efficient, because the Web page that I was directed to ended up being vastly more interesting than the material I was originally trying to track down. Anyway, as any self-respecting mathematical biologist knows by now, d'Arcy Thompson's greatest contribution to theoretical biology was in the last chapter of his magisterial treatise *On Growth and Form,* in which he put forth a theory of biological transformations whereby one might compare the shapes of living things. What popped up on my screen that morning was nothing less than an account of what d'Arcy Thompson would have done if he had had a computer on his worktable instead of a stack of graph paper, a ruler, and a compass.

D'Arcy Thompson held a professorial chair in St. Andrews and Dundee in Scotland for the amazing period of 64 years, a record for tenure unlikely ever to be broken. Although he would write more than 300 scientific articles and books, Thompson's reputation is based primarily upon his attempts to reduce biological phenomena to mathematics in *On Growth and Form*. There he claimed that much about animals and plants could be understood by the laws of physics, as mirrored in the structures and patterns of mathematics. By this, Thompson displayed his essentially anti-Darwinist beliefs, at least in the sense that he would most certainly have disagreed with Theodosius Dobzhansky's well-known remark that, "nothing in biology makes any sense except in the light of evolution." Clearly, Thompson felt that a *lot* of biology made perfectly good sense quite outside the bounds of the principle of natural selection. Besides his academic renown, Thompson seems also to have acquired a bit of a local reputation as a mild eccentric, and older folks in St. Andrews can still recall seeing him strolling about town with a parrot on his shoulder.

The novel idea Thompson set forth in *On Growth and Form* was to show how mathematical functions could be applied to the shapes of one organism to continuously transform it into other, physically similar organisms. A famous example from his book is shown in Figure 5.4, in which a continuous squeezing and stretching of a rectangular Cartesian grid transforms the fish species *Scarus sp.* on the left to the species *Pomacanthus* on the right. Thompson used this same idea to show how to alter pictures of baboon skulls into skulls of other primates, as well as demonstrating how corresponding bones like the shoulder blades are related in different species.

To a topologist, the fact that the fish species *Scarus sp.* can be continuously transformed into the species *Pomacanthus* is entirely

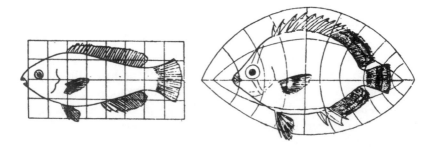

**Figure 5.4**   Two species of fish related by a continuous transformation.

unremarkable. This is because every organism is a closed surface, topologically speaking, and it has been known for many years what kinds of surfaces can and cannot be mapped to one another in various ways. Speaking very loosely, two surfaces are equivalent in this sense if and only if they have the same number of holes. Thus, considering the fact that just about all higher animals have the same number of holes for the digestive tract, ears, nostrils, and eyes, Thompson's work on biological transformations only confirms the topological fact that there exists a continuous transformation that will warp and twist the form of virtually any animal into any other one. So at first glance it doesn't seem as if there is too much meat in Thompson's idea. A world in which aardvarks and zebras are indistinguishable hardly seems to hold much promise for shedding light on the processes of embryology or evolution. But wait!

Simply knowing that there is *some* transformation deforming an aardvark into a zebra is a far different matter from knowing precisely *which* transformation(s) do the job. If there's anything to be said for d'Arcy's view that the basic processes of evolution and development can be understood mathematically via biological transformations, then that something resides in the precise nature of the transformation. It is to address exactly this point that I'm sure d'Arcy would have used a computer had such gadgets been available a century earlier. Fortunately, two of his successors on the faculty of the University of St. Andrews, John J. O'Connor and Edmund F. Robertson, took up this challenge and have created a program for studying the precise analytic form of these Thompson transformations.

The program written by O'Connor and Robertson allows users to alter pictures in real time by varying parameters in mathematical functions describing the transformations, thus seeing the picture change before their very eyes from, say, one fish to another. Figure 5.5 shows the program's user interface corresponding to another of d'Arcy Thompson's famous fish examples. The picture on the left, the species *Argyropelecus olfersi,* is mapped to the unknown species on the right by the quadratic transformation whose parameters are specified at the bottom of the screen. The reader will see from the figure that in this example the transformation only involves a linear stretching of the $x$ and $y$ axes, since the quadratic terms in the transformation are zero.

Symbolically, the program admits maps of the form $(x, y) \rightarrow \big(p(x, y), q(x, y)\big)$, which take a point $(x, y)$ of the plane on the left to

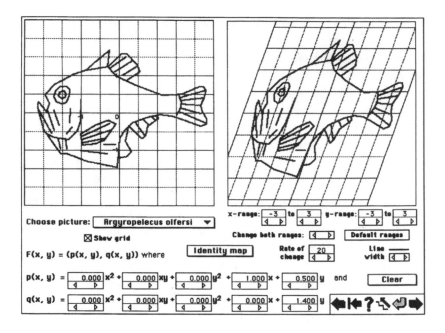

**Figure 5.5** View of the user interface for the Thompson transformation program.

the point $(p, q)$ in the plane on the right, where $p$ and $q$ are quadratic functions in the two variables $x$ and $y$. As a result, there are a total of 10 parameters that can be varied in the process of transforming the source form on the left to any given target form on the right.

Thus, with the O'Connor–Robertson program one can actually twist the 10 "knobs" independently, which corresponds to varying the 10 parameters in $p$ and $q$ until a particular target form is generated. Although Thompson used a wide variety of transformations, O'Connor and Robertson have discovered that most of his effects can be achieved by quadratic maps of the above sort. Consequently, by finding the parameters defining a particular quadratic map leading from one known species to another, we can hope to gain insight into the physical and evolutionary forces acting on different species. Alternately, we can also use the program to study intermediate forms like that of Figure 5.5, which have never before been observed and possibly never even existed. Such novel forms are interesting as objects that *might* exist sometime in the future. They are also of interest as objects of investigation to discover what features they have that prevented their appearance in the past.

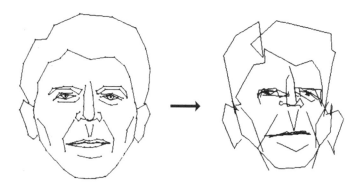

**Figure 5.6**   A computer caricature of Ronald Reagan.

The problem with these kinds of continuous deformations is that nothing essentially new ever turns up; by definition, there can be no *discontinuous* jumps from one species to an entirely different one via a sequence of continuous transformations. For this type of speciation to take place, we need *singularities* in the families of continuous transformations. Although there is no room to discuss the matter further here, let me note in passing that the mathematical theory of catastrophes was developed around 30 years ago to deal with precisely this situation. More information on this line of research is available in the volumes cited in the references.

One set of images used by Thompson in his book was taken from work on facial angles by the artist Albrecht Dürer. These experiments by Thompson on continuous distortions of such angles led to a wide variety of faces, calling to mind a program written a few years ago by Susan Brennan of the Hewlett-Packard Laboratories. Basically, her program is designed to continuously deform a given face in order to create a caricature of it. Serving to motivate the development of this program was the question of how humans so quickly recognize faces, even when only a few features are seen under adverse viewing conditions. Brennan hoped that her program could be used as a tool for investigating just which features people focus upon in this remarkable pattern-recognition process. The details of how the program works have been described elsewhere, but the principle is simplicity itself: Compare the target face with an "average" face, and then scale up those features that differ the most from the average face. An example of how the program caricaturizes former President Ronald Reagan is shown in Figure 5.6.

**Figure 5.7**  A transitional sequence in face space.

Brennan has described her program as a fast way of exploring what she calls "face space." The 186 coordinates she uses to describe a target face can be thought of as the coordinates of a single point in a space of dimension 186. Because every face is a point in this same space, any two faces can be connected by a straight line in face space, each point along the line representing proportional changes in every coordinate value. The distance between any two points then serves as a measure of how similar the faces are. With d'Arcy Thompson in mind, we can also ask the computer to generate a transitional sequence from one face to another. Such a transition sequence from Elizabeth Taylor as Cleopatra to John F. Kennedy is shown in Figure 5.7.

A key factor motivating the work of both d'Arcy Thompson and Susan Brennan is to uncover *invariants* of the transformations from one form to another. If these transformations have any biological or psychological content at all, presumably it lies in telling us what it is *exactly* that allows us to recognize the caricatured version of Ronald Reagan as Ronald Reagan and not, say, JFK, or what *exactly* are the characteristics shared by the fish species *Scarus sp.* and *Pomacanthus*. Only a deeper understanding of such invariants will enable us to unlock the complexity of living things, by showing us the features that remain unchanged as we pass from one species to another along a smooth, evolutionary pathway. As d'Arcy himself put it, "I know that in the study of material things number, order, and position are the threefold clue to exact knowledge; and that these three, in the mathematician's hands, furnish the first outlines for a sketch of the Universe."

From these considerations, I think it is plain to see that scientists like d'Arcy Thompson would have welcomed the computer as an ally in their quest to unlock the secrets of nature and humans. However, in order for complexity theory as embodied in these would-be worlds to have an impact in the real world of science, it will take more than the kind of empirical observations provided by the kinds of models we have been discussing in this volume. It will take a workable *theory* of complex systems. Let us conclude our story with an account of what's needed to put such a theory in place.

## Toward a Theory of the Complex

In chapter 3, we gave a brief account of the El Farol Problem, in which Irishmen in Santa Fe try to decide whether or not to go to the bar to listen to Irish music on Thursday evenings. Although I doubt seriously that any real Irishman like Brian Arthur would be deterred in the least from a visit to the local pub simply by the presence of others of like mind, let's continue the fantasy outlined in the El Farol Problem as a way of encapsulating a large fraction of the problems encountered in the would-be worlds considered in this book.

Recall that the problem faced by the Santa Fe Irishmen was to use their currently best rule to estimate how many music lovers would appear at El Farol in the coming week. On the basis of this prediction, each individual then chose to go to the bar or stay home, with the total number attending being reported to everyone in the following week. At that time, each Irishman revised his or her set of predictors, using the most accurate predictor to estimate the attendance in the coming week. And so it goes, week after week, until the Irishmen got tired of the music and/or the drinking or, what is far more likely (and actually happened), the band of Irish musicians moves on to greener pastures. It is my belief that the key components forming the El Farol Problem are exactly the key components in each and every one of the would-be worlds discussed in the last chapter, and that a decent mathematical formalism to describe and analyze the El Farol Problem would go a long way toward the creation of a viable theory of complex, adaptive systems. So let's look at what these key components are.

***Medium-Sized Number of Agents.*** In the El Farol Problem we have postulated 100 Irishmen, each of whom acts independently in deciding to go or not go to the bar on Thursday evening. In contrast to simple systems—like superpower conflicts, which tend to involve a small number of interacting agents—or large systems—like galaxies or containers of gas, which have a large enough collection of agents that we can use statistical means to study them—complex systems involve what we might call a medium-sized number of agents. Just like Goldilocks's porridge, which was not too hot and not too cold, complex systems have a number of agents that is not too small and not too big, but just right to create interesting patterns of behavior.

*Intelligent and Adaptive.* Not only are there a medium number of agents, these agents are intelligent and adaptive. This means that they make decisions on the basis of rules, and that they are ready to modify the rules they use on the basis of new information that becomes available. Moreover, the agents are able to generate new rules that have never before been used, rather than being hemmed in by having to choose from a set of preselected rules for action. This means that an ecology of rules emerges, one that continues to evolve during the course of the process.

*Local Information.* In our would-be worlds, no single agent has access to what *all* the other agents are doing. At most, each agent gets information from a relatively small subset of the set of agents, and processes this "local" information to come to a decision as to how he or she will act. In the El Farol Problem, for instance, the local information is as local as it can be, because each Irishman knows only what he or she is doing; none have information about the actions taken by any other agent in the system. This is an extreme case, however, and in most would-be worlds the agents are more like drivers in a transport system or traders in a market, each of whom has information about what a few of the other drivers or traders are doing.

So these are the components of almost all complex systems like the El Farol situation—a medium-sized number of intelligent, adaptive agents interacting on the basis of local information. At present, there appears to be no known mathematical structures within which we can comfortably accommodate a description of the El Farol Problem. This suggests a situation completely analogous to that faced by gamblers in the seventeenth century, who sought a rational way to divide the stakes in a game of dice when the game had to be terminated prematurely (probably by the appearance of the police or, perhaps, the gamblers' wives). The description and analysis of that very definite real-world problem led Fermat and Pascal to the creation of a mathematical formalism we now call probability theory. At present, complex-system theory still awaits its

Pascal and Fermat. The mathematical concepts and methods currently available were developed, by and large, to describe systems composed of material objects like planets and atoms. But as philosopher George Gilder has noted, "'The central event of the twentieth century is the overthrow of matter. In technology, economics, and the politics of nations, wealth in the form of physical resources is steadily declining in value and significance. The powers of mind are everywhere ascendant over the brute force of things." It is the development of a proper theory of complex systems that will be the capstone to this transition from the material to the informational.

# References

## Chapter 1 (Reality Bytes)

**Up and Down the Electronic Gridiron**

For football fanatics, the *Football Pro '95* is published by Sierra Online, Box 53250, Bellevue, WA 98015–3250. As computer sports simulations go, this is about as good as things get. For those whose baseball interest runs more to the field than to the bargaining table or courtroom, let me recommend the similarly detailed *Baseball '94* (the last season as of this writing in which real players were on the field instead of lawyers).

For those of a statistical bent, it may be of interest to know that in my experiments the standard deviation of the 54 victories by the 49ers was 3.93 points, whereas it was 5.49 points for the 46 wins by the Chargers. So using 3 standard deviation for a 95% confidence level, we find that the 49er wins should fall between $6.67 \pm 3.93$ points, while Chargers' victories should be in the range $7.15 \pm 5.49$ points—both a far cry from the 19-point spread offered by Las Vegas. As the actual final score was 49 to 26—a 23-point margin—we see what an anomaly, statistically speaking, this game was.

For an account of some of the philosophical—and practical—differences between detailed microworld simulations and mesoworld models, see the essay

Lane, D. "Models and Aphorisms." *Complexity,* 1:2 (1995), 9–13.

**Distant Suns and Planetary Nebula**

A detailed account of Dole's simulated planetary systems is given in

Dole, S. *Habitable Planets for Man,* 2d edition. New York: Elsevier, 1970.

Since these pioneering experiments, a number of other investigators have found similar results. For a discussion of these results within the context of the search for extraterrestrial intelligences, see chapter 6 of

Casti, J. *Paradigms Lost.* New York: Morrow, 1989 (paperback edition: Avon Books, New York, 1990).

**A Gallery of Models**

A good account of the various types of models discussed here is found in

Barbour, I. *Myths, Models, and Paradigms.* New York: Harper & Row, 1974.

**Models for All Occasions**

An introductory account of the problems of prediction and explanation in modern science is found in

Casti, J. *Searching for Certainty.* New York: Morrow, 1991 (paperback edition: Quill Books, New York, 1992).

**The Philosopher's Stones**

The story of Trurl and the evil King Excelsius is but one of several such tales, each addressing one or another philosophical conundrums surrounding the relationship between humans and machines, spun by Stanislaw Lem in

Lem, S. *The Cyberiad.* New York: Seabury Press, 1974.

The quote in the text is found in the story, "The Seventh Sally, or Why Trurl's Perfection Led to No Good," pp. 162–163.

**Semper Fidelis?**

Some of the first-rate discussions of the various problems surrounding artwork, aesthetics, forgeries, creativity and the matter of values are

Battin, M., J. Fisher, R. Moore, and A. Silvers. *Puzzles About Art.* New York: St. Martin's Press, 1989.

*The Forger's Art.* D. Dutton, ed. Berkeley, CA: University of California Press, 1983.

Goodman, N. *Languages of Art.* Indianapolis, IN: Hackett Publishing Co., 1976.

### The Art of the Model

Discussions of the properties characterizing "good" scientific models are found in many volumes on the philosophy of science. Among my favorites are

Bechtel, W. *Philosophy of Science: An Overview for Cognitive Science.* Hillsdale, NJ: Lawrence Erlbaum, 1988.

*Introductory Readings in the Philosophy of Science.* E. Klemke, R. Hollinger and A. Kline, eds. Buffalo, NY: Prometheus Books, 1988.

*Philosophy and Science.* F. Mosedale, ed. Englewood Cliffs, NJ: Prentice Hall, 1979.

### "Hypotheticality"

The problem of global warming must have been described in literally thousands of places by now. One that covers the territory reasonably well is

Schneider, S. *Global Warming.* San Francisco: Sierra Club Books, 1989.

A first-rate volume that considers the entire issue of cyclical behavior in climatic affairs is

Burroughs, W. *Weather Cycles: Real or Imaginary?* Cambridge: Cambridge University Press, 1992.

For accounts of genetic engineering, fiscal policy, and/or AIDS vaccines, I refer the interested reader to his or her favorite daily newspaper or TV newscasters.

### From Real Space to Cyberspace

For those who have spent the last decade in outer space or in Rip-Van-Winkle-like hibernation, here are two of the more reasoned discussions of the virtual-reality phenomenon.

Heim, M. *The Metaphysics of Virtual Reality.* New York: Oxford University Press, 1993.

Woolley, B. *Virtual Worlds.* Oxford: Basil Blackwell, 1992.

# Chapter 2 (Pictures as Programs)

### That's Life?

An excellent layman's discussion of the whole artificial-life movement, including a detailed account of Langton's intellectual "rebirth," is given in

Levy, S. *Artificial Life.* New York: Pantheon, 1992.

A full-scale account of the work by Dawkins and Pickover in the creation of Biomorphland and its inhabitants can be found in

Dawkins, R. "The Evolution of Evolvability." In *Artificial Life,* C. Langton, ed. Reading, MA: Addison-Wesley, 1989, pp. 201–220.

Pickover, C. "Biomorphs: Computer Displays of Biological Forms Generated from Mathematical Feedback Loops." *Computer Graphics Forum,* 5:4 (1987), 313–316.

Work by Lindenmayer, Prusinkiewicz, and their coworkers on L-systems and their use in generating artificial plants is described in glorious color in

Prusinkiewicz, P., and A. Lindenmayer. *The Algorithmic Beauty of Plants.* New York: Springer, 1990.

**The Genesis Machine**

The Turing machine, ASCII code, and artificial intelligence are discussed in countless books and articles. One source for layman, where all three are considered, is

Casti, J. *Paradigms Lost.* New York: Morrow, 1989 (paperback edition: Avon Books, New York, 1990).

The adventuresome reader will find many additional technical items cited in the reference section of this work.

A rather hilarious collection of conversations, poems, and just plain random ramblings by *Racter* is contained in the book

*The Policeman's Beard is Half Constructed.* New York: Warner Books, 1984.

An introduction to how *Racter* actually works is found in

Dewdney, A. "Conversations with *Racter.*" In *The Armchair Universe.* New York: W. H. Freeman, 1988, pp. 79–88.

**The Electronic Abacus**

A complete account of the Four-Color Conjecture and its solution is available in

Saaty, T., and P. Kainen. *The Four-Color Problem.* New York: McGraw-Hill, 1977 (Dover reprint, New York, 1986).

For a consideration of the philosophical import of the computer solution of the Four-Color Problem, see

Tymoczko, T. "The Four-Color Problem and Its Philosophical Significance." *Journal of Philosophy,* 76:2 (1979), 57–83.

The question of how proteins fold in three dimensions in order to carry out various functions is a prime staple of molecular biology, and it would certainly be of great theoretical and practical importance to be able to predict the folding pattern on the basis of the linear string of amino acids. For an account of the protein-folding problem from a biological point of view, the reader should consult

Stryer, L. *The Molecular Design of Life.* New York: W. H. Freeman, 1989.

For a proof that the mathematical version of protein folding is a computationally "hard" problem, see

Fraenkel, A. "Complexity of Protein Folding." *Bulletin of Mathematical Biology,* 55:6 (1993), 1199–1210.

The work by Karplus and his colleagues is reported in

Šali, A., E. Shakhnovich, and M. Karplus. "How Does a Protein Fold?" *Nature,* 369 (19 May 1994), 248–251.

For an introduction to computational weather forecasting, see chapter 2 of

Casti, J. *Searching for Certainty.* New York: Morrow, 1991 (paperback edition: Quill Books, New York, 1992).

A rather more technical treatment of the entire issue of numerical simulation of the atmosphere is found in

Daley, R. *Atmospheric Data Analysis.* New York: Cambridge University Press, 1991.

**The Creative Computer**

Chomsky's work has been chronicled in so many places that it's hard to know where to start by way of giving pointers to the literature. One good place for layman, though, is chapter 4 of

Casti, J. *Paradigms Lost.* New York: Morrow, 1989 (paperback edition: Avon Books, New York, 1990).

Those who doubt Chomsky's claim that language acquisition is an innate property of the human brain will find ammunition for their position in the article

> Sampson, G. "Language Acquisition: Growth or Learning?" *Philosophical Papers,* 18:3 (1989), 203–240.

An excellent recent account of the debate Chomsky's work has engendered in the linguistics community is

> Harris, R. *The Linguistic Wars.* New York: Oxford University Press, 1993.

The discussion in the text of tonal music and Schoenberg's work follows that given in the fascinating volume

> Holtzman, S. *Digital Mantras.* Cambridge, MA: MIT Press, 1994.

This book is notable not only for its treatment of music as a generative art form, but also for showing how the art of Wassily Kandinsky and Harold Cohen falls under the same kind of formal constraints.

Karl Sims's ideas about how to employ evolution with LISP programs to create artistic forms was first presented in

> Sims, K. "Artificial Evolution for Computer Graphics." *Computer Graphics,* 25:4 (1991), 319–328.

**Worlds in Silico**

Chris Crawford's detailed account of why and how he created *Balance of Power* makes for about as good a discussion as one can find of the ins and outs of computer game design. It is given in

> Crawford, C. *Balance of Power.* Redmond, WA: Microsoft Press, 1986.

For an introduction to both classical and avant garde ideas in finance, see chapter 4 of

> Casti, J. *Searching for Certainty.* New York: Morrow, 1991 (paperback edition: Quill Books, New York, 1992).

The stock market simulation by Mssrs. Arthur, Holland et al. is described in detail in

> Arthur, W. Brian, J. Holland, B. LeBaron, R. Palmer, and P. Tayler. "Asset Pricing under Inductive Reasoning in an Artificial Stock Market." Santa Fe Institute Working Paper, to appear 1996.

A very similar exercise in bottom-up simulation involving the German stock exchange is reported in

Rieck, C. "Evolutionary Simulation of Asset Trading Strategies." In *Many-Agent Simulation and Artificial Life,* E. Hillebrand and J. Stender, eds. Amsterdam: IOS Press, 1994, pp. 112–136.

# Chapter 3 (The Science of Surprise)

## Problems and Paradoxes

A general reference for all the material of this chapter is the popular volume

Casti, J. *Complexification.* New York: HarperCollins, 1994 (paperback edition: Harper Perennial, New York, 1995).

A mathematical treatment of the Bouncer showing its chaotic nature is given in the article

Berry, M. "The Unpredictable Bouncing Rotator: A Chaology Tutorial Machine." In *Dynamical Systems: A Renewal of Mechanism,* S. Diner et al., eds. Singapore: World Scientific, 1986, pp. 3–12.

The example showing the counterintuitive nature of the developing economy is adapted from

Bristol, E. "The Counterintuitive Behavior of Complex Systems." *IEEE Systems, Man & Cybernetics Newsletter,* March 1975.

## The Fingerprints of the Complex

A first-rate account of logical paradoxes like the Liar is given in the little volume

Leiber, J. *Paradoxes.* London: Duckworth, 1993.

A more extensive discussion of paradoxes of all sorts—logical and visual—including the impossible staircase, is found in

Falletta, N. *The Paradoxicon.* New York: Doubleday, 1983.

A good source for material on positive feedbacks in the economy is the collection of papers

Arthur, W. B. *Increasing Returns and Path Dependence in the Economy.* Ann Arbor, MI: University of Michigan Press, 1994.

The question of whether the human mind operates exclusively on the basis of rules lies at the heart of the "mind-as-machine" debate. A reasonably up-to-date account of the various competing theories and philosophies on the question is given in chapter 5 of

Casti, J. *Paradigms Lost.* New York: Morrow, 1989 (paperback edition: Avon Books, New York, 1990).

A good summary of the history of the $N$-Body Problem and its current mathematical status, including the details of Xia's solution discussed in the text, is given in

Saari, D., and Z. Xia. "Off to Infinity in Finite Time." *Notices of the American Mathematical Society,* 42 (1995), 538–546.

Deborah Gordon's work on harvester ant colonies and the emergence of work assignments within the colony is recounted in

Gordon, D. "The Development of Organization in an Ant Colony." *American Scientist,* 83 (January–February 1995), 50–57.

### Illusions of the Mind

Arrow's Impossibility Theorem is discussed in just about every book on decision making. A good place to get more information about how this result arises in various choice situations is

MacKay, A. *Arrow's Theorem: The Paradox of Social Choice.* New Haven, CT: Yale University Press, 1980.

Nomination paradoxes can arise wherever decisions are made by voting. For a more detailed account of such sitations, see

Brams, S. *Paradoxes in Politics.* New York: The Free Press, 1976.

### A Little Can Be a Lot

The treatment of the rise and fall of the Roman Empire follows that in

Woodcock, A., and M. Davis. *Catastrophe Theory.* New York: Dutton, 1978.

Detailed accounts of chaotic processes are found in many places. Good sources for the layman are

Casti, J. *Reality Rules–I.* New York: Wiley, 1992.

Gleick, J. *Chaos.* New York: Viking, 1987.

An introductory account of weather forecasting and chaos is given in the article

Palmer, T. "A Weather Eye on Unpredictability." In *The* New Scientist *Guide to Chaos,* N. Hall, ed. London: Penguin, 1992, pp. 69–81.

## The Rules of the Game

The El Farol Problem was first presented in

Arthur, W. B. "Inductive Reasoning and Bounded Rationality." *American Economic Association Papers and Proceedings,* 84 (1994), 406–411. (Also available on the World Wide Web at URL http://www.santafe.edu/arthur/.)

Sonnenschein's result on the complexity of economic processes is discussed in the broader context of how such counterintuitive processes permeate the social sciences in the article

Saari, D. "Mathematical Complexity of Simple Economics." *Notices of the American Mathematical Society,* 42 (1995), 222–230.

An introduction to the ideas underlying the work of Gödel, Turing, and Chaitin is found in chapter 7 of

Casti, J. *Searching for Certainty.* New York: Morrow, 1991 (paperback edition: Quill Books, New York, 1992).

## The Connections That Count

Kauffman nets are discussed in detail in

Kauffman, S. "Antichaos and Adaptation." *Scientific American,* 265 (August 1991), 78–84.

Kauffman, S. *Origins of Order.* New York: Oxford University Press, 1992.

A fascinating introduction to the use of algebraic topology in the treatment of not only chess, but also questions ranging from road-traffic flow to the structure of cities and the work of the artist Piet Mondrian is given in the volume

Atkin, R. *Mathematical Structure in Human Affairs.* London: Heinemann, 1974.

## The Laws of Emergence

The extraordinary history of the Central Limit Theorem is chronicled in

Adams, W. *The Life and Times of the Central Limit Theorem.* New York: Kaedomon, 1974.

For an account of Zipf's Law in all its many linguistic manifestations, see the works

Flam, F. "Hints of a Language in Junk DNA." *Science,* 266 (25 November 1994), 1320.

Nicolis, J. *Chaos and Information Processing.* Singapore: World Scientific, 1991.

Mandelbrot, B. "On the Theory of Word Frequencies and on Related Markovian Models of Discourse." In *Structures of Language and Its Mathematical Aspects,* R. Jacobson, ed. New York: American Mathematical Society, 1961.

Schroeder, M. *Fractals, Chaos, Power Laws.* New York: W. H. Freeman, 1991.

Zipf, G. *Human Behavior and the Principle of Least Effort.* Cambridge, MA: Addison-Wesley, 1949.

Li, W. "Random Texts Exhibit Zipf's-Law-like Word Frequency Distribution." *IEEE Transactions on Information Theory,* IT–38:6 (1992), 1842–1845.

# Chapter 4 (Artificial Worlds)

## A Louisiana Saturday Afternoon

Unfortunately, at the moment there is no publicly available account of the world of *TRANSIMS.* Hopefully, this deplorable situation will be rectified shortly. However, a glimpse into some of the theoretical issues underlying this simulation is available in the report

Rasmussen, S., and C. Barrett. "Elements of a Theory of Simulation." Santa Fe Institute, Working Paper 95–04–040, 1995.

## Microworlds, Macrobehaviors

*Corewars* was introduced to the world by A. K. Dewdney in the pages of *Scientific American* magazine. The articles explaining the game, as well as after-the-fact comments and observations, are found in the volumes

Dewdney, A. *The Armchair Universe.* New York: W. H. Freeman, 1988 (section 7, pp. 275–309).

Dewdney, A. *The Magic Machine.* New York: W. H. Freeman, 1990 (chapters 27–28).

For those who would like to try their hand at creating a warrior to do battle in the Corewar Colosseum, the *Corewar* simulator for use under Windows is available on the Internet via anonymous file transfer from `ftp.csua.berkeley.edu` under directory `/pub/corewar/systems`.

**How Life Learns to Live**

Neural nets are perhaps the most actively pursued area of study nowadays in the world of applied computing. Accordingly, booksellers shelves bulge with the volume of volumes appearing on the topic. One of the best no-nonsense introductions is

Stanley, J. *Introduction to Neural Networks.* Pasadena, CA: California Scientific Software, 1990.

Unfortunately, this volume comes bundled with the software package *Brain-maker,* and so may be very hard to find (without, of course, purchase of the software). Neural network research sometimes masquerades under the grandiose rubric "parallel distributed processing" (PDP). A pioneering work in this area containing contributions from almost all the leading workers in the field is

*Parallel Distributed Processing,* Vols I and II. J. McClelland, D. Rumel-hart et al., eds. Cambridge, MA: MIT Press, 1986.

This encyclopedic treatise on PDP also has an accompanying tutorial volume, containing software suitable for experimentation with many of the methods and algorithms described in the main book itself. Other introductory accounts of neural networks include

Johnson, J., and H. Picton. *Mechatronics: Designing Intelligent Ma-chines.* Oxford: Butterworth-Heinemann, 1995.

Picton, H. *Introduction to Neural Networks.* Basingstoke, UK: Macmil-lan, 1994.

Lipmann, R. "An Introduction to Computing with Neural Nets." *IEEE ASSP Magazine,* (1987), 4–22.

Johnson, J. "An Introduction to Neural Networks." The Open University, Milton Keynes, UK, preprint, August 1995.

The question of whether a machine can be made to think just like a human has captured the imagination almost from the very dawning of the computer age. As one might expect, the literature on the topic is voluminous. The interested reader is referred to the following sources on the subject.

*Thinking Computers & Virtual Persons.* E. Dietrich, ed. San Diego, CA: Academic Press, 1994.

*The Artificial Intelligence Debate.* S. Graubard, ed. Cambridge, MA: MIT Press, 1988.

Johnson, G. *Machinery of the Mind.* New York: Times Books, 1986.

McCorduck, P. *Machines Who Think.* San Francisco: W. H. Freeman, 1979.

Books and articles on evolutionary processes probably constitute the single largest fraction of material published in the scientific world. There seems to be no facet of the topic that has not been explored *ad nauseum,* yet, no publisher's catalogue of current offerings is complete without at least one more entry to the ongoing, seemingly eternal, debate as to whether Darwin got it right or wrong or somewhere in between. So let me content myself with listing just a few of my favorite entries in this evolutionary sweepstakes, chosen for their historical, scientific, philosophical and literary qualities.

Dawkins, R. *River Out of Eden.* New York: Basic Books, 1995.

Haldane, J. B. S. *The Causes of Evolution.* Princeton, NJ: Princeton University Press, 1990.

Maynard Smith, J. *The Theory of Evolution.* Cambridge: Cambridge University Press, 1993.

Ruse, M. *The Darwinian Paradigm.* London: Routledge, 1989.

*Darwin,* 2d ed. P. Appleman, ed. New York: Norton, 1979.

The idea of mimicking the evolutionary processes of nature in a computing machine to "grow" solutions to problems was pioneered by John Holland. For his original account of this work done in the 1960s, see the volume

Holland, J. *Adaptation in Natural and Artificial Systems.* Ann Arbor, MI: University of Michigan Press, 1975.

A more up-to-date discussion is found in

Holland, J. *Hidden Order.* Reading, MA: Addison-Wesley, 1995.

Other introductory accounts of genetic algorithms and their uses are

Michalewicz, Z. *Genetic Algorithms + Data Structures = Evolution Programs,* 2d ed. Berlin: Springer, 1994.

Mitchell, M. "Genetic Algorithms: An Overview." *Complexity,* 1 (1995), 31–39.

Holland, J. "Genetic Algorithms." *Scientific American,* (July 1992), 66–72.

Denning, P. "Genetic Algorithms." *American Scientist,* 80 (January–February 1992), 12–14.

Mitchell, M., and S. Forrest. "Genetic Algorithms and Artificial Life." *Artificial Life,* 1 (1994), 267–289.

## *Tierra*

Tom Ray's work has been chronicled in many places. Probably the best single account is Ray's own article

Ray, T. "An Approach to the Synthesis of Life," in *Artificial Life–II,* C. Langton et al., eds. Redwood City, CA: Addison-Wesley, 1991, pp. 371–408.

Other introductory accounts by Ray include

Ray, T. "An Evolutionary Approach to Synthetic Biology: Zen and the Art of Creating Life." *Artificial Life,* 1 (1994), 195–226.

Ray, T. "Evolution, Complexity, Entropy, and Artificial Reality." *Physica D,* 75 (1994), 239–263.

For those interested in exploring *Tierra* on their own, the program and documentation is available on the Internet. The relevant URL is http://vrml.arc.org/tierra. An account of Ray's idea to distribute *Tierra* on the Internet so as to give the program a much broader playing field is discussed in

Ray, T. "A Proposal to Create a Network-Wide Biodiversity Reserve for Digital Organisms." (Available on ftp site life.slhs.udel.edu in the file reserves.tex.)

For a journalistic account of Pargellis's work on the creation of living objects without benefit of an Ancestor organism, see

Guinnessy, P. " 'Life' Crawls Out of the Digital Soup." *New Scientist,* (April 13, 1996), 16.

### Societies in Silicon

The work by Epstein and Axtell on creating artificial societies is covered in great detail in their monograph

Epstein, J., and R. Axtell. *Growing Artificial Societies: Social Science from the Bottom-Up.* Princeton, NJ: Princeton University Press, 1996.

The potential for would-be worlds like the *CompuTerrarium* for illuminating issues in economics is explored in the article

Lane, D. "Artificial Worlds and Economics, Parts I and II." *J. Evolutionary Economics,* 3 (1993), 89–107 and 177–197.

### Building Silicon Worlds

John Miller's work on testing simulation models is reported in

Miller, J. "Active Nonlinear Tests (ANTs) of Complex Simulation Models." Santa Fe Institute Working Paper, WP–96–03–011, March 1996.

The *SWARM* system is described in the paper

Langton, C., N. Minar, and R. Burkhart. "'The Swarm Simulation System: A Tool for Studying Complex Systems." Santa Fe Institute preprint, March 1995.

Those wishing a copy of the *SWARM* software can download it from the World Wide Web at URL http://www.santafe.edu/projects/swarm/.

# Chapter 5 (Reality of the Virtual)

### Inside Information

The first explicit consideration of what we would now call endophysics appears to be in the work of James Clerk Maxwell. For a discussion of Maxwell's ideas, as well as for an account of subsequent developments in the endophysics business, see the paper

Rössler, O. "Endophysics." In *Real Brains, Artificial Minds,* J. Casti and A. Karlqvist, eds. New York: Elsevier, 1987, pp. 25–46.

The term "endophysics" appears to have been introduced by the quantum physicist David Finkelstein, who also gave an early example of a finite-state machine (computer) whose internally evaluated state is different from that existing exo-physically. For an account of this device, see the article

Finkelstein, D., and S. Finkelstein. "Computational Complementarity." *Intl. J. Theoretical Physics,* 22 (1983), 753–779.

A recent summary of many of the ideas underpinning the endo/exo distinction is given in

*Inside versus Outside.* H. Atmanspacher and G. Dalenoort, eds. Berlin: Springer, 1994.

**The Moral of the Model**

The aphorisms discussed in this section are treated in more detail in

Lane, D. "Models and Aphorisms." *Complexity,* 1:2 (1995), 9–13.

**Limits to Scientific Knowledge**

The issue of limits to what we can hope to know by following a set of scientific rules has recently become a topic of concern in many quarters. The treatment given here follows that in

Casti, J. "Limits to Scientific Knowledge." *Scientific American,* to appear October 1996.

Other accounts include

*Limits of Predictability.* Yu. Kravtsov, ed. Berlin: Springer, 1994.

Jackson, E. Atlee. "No Provable Limits to 'Scientific Knowledge'." *Complexity,* 1:2 (1995), 14–17.

Geroch, R., and J. Hartle. "Computability and Physical Theories." *Foundations of Physics,* 16:6 (1986), 533–550.

Traub, J. "What is Scientifically Knowable?" In *CMU Computer Science,* R. Rashid, ed. Reading, MA: Addison-Wesley, 1991, pp. 489–503.

"On Limits." J. Casti and J. Traub, eds. Working Paper WP–94–10–056, Santa Fe Institute, 1994.

**Computation *Über Alles***

The discussion of what d'Arcy Thompson might have done with a computer at his disposal follows that in

Casti, J. "If D'Arcy Had Only Had a Computer." *Complexity,* 1:3 (1995), 5–8.

Those wishing to have a look at the program by O'Connor and Robertson should consult the World Wide Web at URL http://www-groups.dcs.st-and.ac.uk/~history/Miscellaneous/darcy.html.

For an account of d'Arcy Thompson's work, see

Thompson, d'Arcy W. *On Growth and Form.* Cambridge: Cambridge University Press, 1917 (Dover reprint edition, 1992).

Rosen, R. "Dynamical Similarity and the Theory of Biological Transformations." *Bulletin of Mathematical Biology,* 40 (1978), 549–579.

Good introductory references to catastrophe theory include

Arnold, V. I. *Catastrophe Theory.* Berlin: Springer, 1984.

Poston, T., and I. Stewart. *Catastrophe Theory and Its Applications.* London: Pitman, 1978.

Saunders, P. *Introduction to Catastrophe Theory.* Cambridge: Cambridge University Press, 1980.

Thom, R. *Structural Stability and Morphogenesis.* Reading, MA: Addison-Wesley, 1975.

Zeeman, E. C. *Catastrophe Theory: Selected Papers, 1972–1977.* Reading, MA: Addison-Wesley, 1977.

Readers wishing to know more about Susan Brennan's work on face shapes should consult

Brennan, S. "Caricature Generation: The Dynamic Exaggeration of Faces by Computer." *Leonardo,* 18:3 (1985), 170–178. (See also the account in A. K. Dewdney, "Facebender," in *The Armchair Universe,* New York: W. H. Freeman, 1988, pp. 89–99.)

# Index